Science with Reason

*Edited by Sue Atkinson
and Marilyn Fleer*

Heinemann
Portsmouth, NH

'I hate teaching science. I don't understand it. I'm no good at it. I just hate every minute of doing it . . . My poor kids, they all hate it too.'

Teacher on a course

This book is dedicated to all teachers who feel like this.

First published in 1995

Heinemann
A Division of Reed Elsevier Inc.
361 Hanover Street
Portsmouth, NH 03801-3912

Offices and agents throughout the world

Copyright © Sue Atkinson and Marilyn Fleer 1995

All rights reserved. No part of this publication may be reproduced or transmitted in any other form or by any means, electronic or mechanical, including photocopy, recording, or any information storage and retrieval system, without permission in writing from the publisher.

The acknowledgements constitute an extension of this copyright notice.

ISBN 0-435-08381-3

Library of Congress Cataloging-in-Publication Data
On file with the Library of Congress.

Published simultaneously in the United States
in 1995 by Heinemann
and in the United Kingdom by
Hodder and Stoughton Educational,
a division of Hodder Headline Plc,
338 Euston Road,
London NW1 3BH

First published 1995
Impression number   10 9 8 7 6 5 4 3 2 1
Year   1999 1998 1997 1996 1995

Typeset by Wearset, Boldon, Tyne and Wear.
Printed in Great Britain for Hodder & Stoughton Educational, a division of Hodder Headline Plc, 338 Euston Road, London NW1 3BH by The Bath Press, Bath.

Contents

About the editors		iv
About the contributors		iv
Acknowledgements		vi
Editors' introduction: What do we mean by 'science with reason'?		vii

SECTION A: THE CHALLENGE OF TEACHING AND LEARNING IN SCIENCE

1. Approaches to teaching and learning in science   2
   Marilyn Fleer

2. Science in early childhood   9
   Maulfry Hayton

3. Science from a child's perspective   15
   Jonathan Osborne

SECTION B: DEVELOPING TEACHING STRATEGIES

4. Children making paper planes   26
   Sue Atkinson

5. Talking it through: young children thinking science   32
   Maulfry Hayton

6. Starting science from talking, listening and questioning   42
   Rosemary Stickland

SECTION C: TEACHERS' STORIES

7. Developing children's understanding of their bodies   48
   Careen Leslie

8. Putting the green back into plants   58
   Vicky Bresnan

9. A moving experience   67
   Marita Corra

10. Straight bits and bendy bits: an exploration of paper bridges   75
    Lindsey Weimers

11. Exploring telephones   82
    Jan Elliot

12. Science from a building site   92
    Sue Atkinson

13. Exploring outer space   100
    Karina Sargeson

14. Helping to save our planet   107
    Karina Sargeson

15. Them bones, them bones are not all dry!   115
    Mary Sofo

16. What is the universe made of?   125
    Jan Elliot

17. Rock week   135
    Jill Jesson

SECTION D: A PRACTICAL APPROACH

18. How can I organise myself?   148
    Sue Atkinson

19. Writing a primary science policy   155
    Graham Peacock

20. Assessment and achievement   160
    Shirley Clarke

21. Getting better at investigations   166
    Robin Smith

22. Conclusions   170
    Sue Atkinson

References	173
Resources List	176
Index	179

About the editors

SUE ATKINSON is a writer and primary teacher and lectures part time at the Institute of Education, London University. Her doctoral research interests included looking at ways to help teachers feel more confident about their work, and the professional development of teachers, and she is currently interested in the links between the ways in which we teach language and the other areas of the curriculum.

MARILYN FLEER is a senior lecturer at the University of Canberra, Australia. She lectures in science and technology education and human development and learning. She has taught children aged 3–8 and has experience with curriculum development for the Curriculum Corporation of Australia, and in Aboriginal Education. She has a Ph.D. in early childhood science education. Currently she is developing early childhood support materials for the National Technology Statement and Profile on behalf of the Curriculum Corporation of Australia.

About the contributors

VICKY BRESNAN is an experienced pre-school teacher who is always looking to improve her teaching. She continually challenges herself by participating in inservice, conferences and a range of teaching projects. Currently she is tutoring part time at the University of Canberra in the Early Childhood degree course and working full time at Duffie Pre-school in Canberra, Australia.

SHIRLEY CLARKE was a primary teacher for ten years and is now INSET co-ordinator for assessment in the INSET department of the Institute of Education, London University. She co-ordinates a wide range of Institute-based and outreach courses and was a part of the CATS development agency for Key Stage 1 SATS. She has written various books and articles on primary maths and primary assessment.

MARITA CORRA is an experienced early childhood teacher who has been very active in science and technology curriculum development in recent years. She is currently tutoring part time at the University of Canberra in the Early Childhood degree course and teaching at the Yarralumla pre-school in Canberra, Australia.

JAN ELLIOT is a senior teacher at Torrens Primary School in Canberra, Australia, although she has taught in many Australian schools. Jan is well known for her innovative teaching in science in Canberra. In 1989 she won a UK–Australia fellowship in science teaching. Jan toured the UK sharing and gaining ideas from other teachers and academics in England.

MAULFRY HAYTON has taught throughout the primary years and has a special love of teaching in the early years. She is currently teaching at Ide First School, Exeter, Devon and during recent years she has taught at the College of St. Mark and St. John, Plymouth and has run INSET for teachers throughout Devon. She has recently completed her M.Ed. on the quality of learning in a High/Scope setting and is involved in a 'Philosophy for Children' project. Her research interests include developmentally appropriate learning and the development of children's schemes.

JILL JESSON is a writer and a primary teacher and is currently studying for an M.Sc. at Sheffield Hallam University. She has written several books on primary science, writes for Scholastic magazines and runs inservice courses for primary teachers in technology.

CAREEN LESLIE is the Director of the Wiradjuri Pre-school Childcare Centre at the University of Canberra. She has worked as an early childhood teacher for ten years in a variety of centres all around Australia. She has also tutored at the University of Canberra in the Early Childhood degree course.

JONATHAN OSBORNE works at the Centre for Educational Studies, King's College London as a lecturer in science education. His main research interest is young children's understanding of science and he had a major involvement in the Science Process and Concept Exploration (SPACE) project, (1988–92) directed by Professor Paul Black and Professor Wynne Harlen. This project led to the production of the Nuffield Primary Science materials (see resources list). He also is interested in explaining science to non-specialists and wrote a book on physics to help secondary teachers.

GRAHAM PEACOCK is a senior lecturer in science education at Sheffield Hallam University. Before that he was a deputy head in a primary school. He has taught in infant, junior and secondary schools. His research interests include finding out about how teachers gain and use science subject knowledge. He has published several books about science for teachers and children.

KARINA SARGESON is a newly graduated teacher. She has taught at Holt Primary School in Canberra, Australia for two years. Her classroom is organised into learning centres where the children work on negotiated learning tasks. Whilst this approach is common in the UK it is quite unusual in Australia.

ROBIN SMITH is a principal lecturer at Sheffield Hallam University with responsibility for primary education, specialising in science. He has taught in schools in London and York. His research interests include teachers' knowledge of science and the nature of scientific investigations.

MARY SOFO is a Special Education Teacher at a catholic primary school in Canberra and she has worked with children with learning difficulties for fifteen years. She has developed her teaching of science in the ways she describes in the book since she noticed that this way of teaching supported the slow learner, helped them to feel worthwhile, raised their self-esteem and increased their motivation. She is also engaged in teaching adults and has presented many conference papers.

ROSEMARY STICKLAND is a primary teacher in Oxfordshire and worked on the National Oracy Project (this was a project over three years with teachers doing research in the classroom into speaking and listening in all areas of the curriculum). Her main interests in teaching are developing children's questioning and their ability to communicate with each other.

LINDSEY WEIMERS is a primary school teacher who also lectures part time at Westminster College, Oxford. She has been involved in research with the National Primary Centre and is at present studying for an M.A. at Oxford Brookes University.

Acknowledgements

We would like to thank the following for their help in the preparation and writing of this book: Karen Baron; Robyn Triglone; Petrie Murcheson; the University of Canberra, both for the use of their resources and the encouragement received from colleagues; Alan Nicol, for his assistance with video taping Jan's teaching in Chapter 11 and Careen's teaching in Chapter 7 – this video made it much easier for these two teachers to write their stories; Maulfry Hayton; Mary Wilkinson and Elizabeth Carruthers, for sharing their research into the ways in which children learn – working with them and their group in Devon on 'emergent' or 'whole' learning was both stimulating and encouraging; Graham Peacock and Robin Smith, who shared some of their work and ideas; and Shirley Clarke, who helped with an earlier draft of the book. We particularly want to thank Jonathan Osborne for his help and insights in the final stages of preparing the typescript.

Thanks also to the many others who have helped over the years in refining some of the ideas included in this book, particularly the different ways in which children can record their work and the use of a science board.

We have tried to cover a broad range of science topics, but we couldn't cover them all. We have chosen stories and chapters that we hope will make you want to say, 'I could do that!' If the book does that, we will be delighted.

Sue Atkinson and Marilyn Fleer, 1995

Editors' introduction: What do we mean by 'Science with reason'?

As editors, we are committed to the view that learning in young children takes place through participation in purposeful activity; and that it needs to give a very high profile to the role of language in conceptual development. Learning is about a partnership between children, teachers and parents. As we see it the learning process includes:

- purposeful activities, in which children can ask their own questions, and solve their problems their way;
- building on what children already know;
- making it clear to children why they are doing certain activities so that they are able to reflect on and evaluate what they do;
- using the children's own ideas, and building on them to enable them to construct meanings;
- encouraging the children to explore their own means of recording;
- using these different means of recording to help us reveal the children's thinking;
- enabling the children to take responsibility for their own learning.

(This process is described more fully in *Mathematics with Reason*, Atkinson, 1992).

The purpose of this book is to relate this approach to learning science. This is what we mean by science with reason. We hope that, through reading this book, you will journey through many exciting and positive experiences in science teaching. We focus not just on recent pedagogical developments in science, but show, through teachers' own stories of life and living processes, materials and their properties, physical processes, and scientific investigations, how these areas can be integrated into children's learning. The stories not only give a picture of what science learning can look like for a wide range of topics, but they provide excellent ideas for starting off each topic. Similarly, a variety of useful, practical ideas are given by teachers in their stories. We hope these will help you to organise teaching and learning in science for your pupils. In short, this book presents a holistic approach to science learning for children, which you can read from start to finish, or dip into like a recipe book.

Our case studies are about children up to the age of eleven and we suggest that what works for a five- or eleven-year-old, will, with some adaptations, probably work for all children (and adults, too). We are looking at a way of learning that has been variously described as the 'whole language approach', 'the emergent approach', the 'developmental approach', the 'whole child approach' and 'a constructivist approach'. As teachers, we find these ways of teaching both exciting and demanding.

We have tried to strike a balance in the book between research, theory and practical examples and to show a pattern – a general way of working – in the teaching and learning of science. We hope that, as teachers, students and parents read, this pattern will begin to emerge. It starts with finding out what the children already know, then asking the children (in various ways) what it is they now want to know more about, and finally working alongside the children to build on their knowledge.

In science, research into children's understandings by Rosalind Driver (1983), the SPACE project in the U.K., and Roger Osborne and other researchers from the University of Waikato in New Zealand, has shown us the importance of listening to children's ideas. This research has demonstrated that, unless we take account of the intuitive ideas held by children, we may never know the real impact our science experiences are having on them. Each teacher's story, therefore, has focused on finding out what children know before they organise science learning for them.

The rest of this introductory chapter offers a synopsis of the material covered in the remainder of the book.

In **section A** we introduce some of the main ideas of the book. In science, the notion of taking account of children's 'intuitive' ideas (if we can call them that) has come into the foreground over the last ten years. Yet not all approaches to teaching in science take account of this research.

Over the years, we have been exposed to an array of teaching techniques for many different areas of the curriculum. Sometimes the approaches are complementary and sometimes they can even be contradictory. Rarely have we had one approach for all curriculum areas. This is not surprising given the diversity of knowledge, skills and differing attitudes in the curriculum areas. There are a number of different approaches which Chapter 1 will describe more fully.

Many of the chapters later in the book describe what we will call an 'interactive' approach to teaching science. Others, such as Chapter 12, are influenced much more by a process approach. It will become apparent that different approaches are needed by teachers at different times and we probably all use some aspects of each approach at some time, depending on what aspect of the curriculum we are working on with the children.

However, we are trying to say in the book that there are some ways of teaching science that we think are very much more effective than others. We base that view on our observations of the outcomes of what the children have learned, what they retain, and the explanations that they give of their understandings after they have worked on an aspect of the curriculum.

In Chapter 3, Jonathan Osborne writes about some of the research findings of the SPACE project. He demonstrates the importance of the role of language in the gradual understanding of scientific concepts and the crucial need within that process for the teacher to try to understand what the child already knows. He points out that one of the reasons that science is so difficult is because our common sense is not very helpful when it comes to understanding some scientific reasoning and this is particularly true for children.

So, the finding out of what the child already knows is a crucial aspect of this approach and this is a repeated theme in the book and an important feature of 'good' science teaching. We are clearly advocating a constructivist approach to teaching science – building on what the child already knows.

Constructivism acknowledges that children actively construct meaning for themselves. The 'intuitive' ideas held will influence how the child constructs meaning and what sense the child will make of the learning experiences we provide. Consequently, we need to plan actively to listen to children, so that we can begin to understand their 'intuitive' ideas.

Chapter 3 also demonstrates the crucial need for the teacher to get children to describe what they are observing as they engage in scientific learning. Here again, language is crucial and Jonathan suggests that it is by encouraging children to be explicit, and by teachers being explicit themselves, that we will be able to encourage children towards developing their scientific understandings. For example, as teachers, we can question children about what exactly they mean as they try to describe what they are doing and what they are observing. This role of the teacher is crucial – totally different from their imagined role where children are let loose to 'discover for themselves'.

In **section B** there are three chapters that focus particularly on the different classroom strategies that teachers can use to encourage their children to talk and question what they do. Finding out what children's ideas are is crucial in teaching science, but this has to be planned for and in this section three teachers discuss how they encourage children to question and develop their ideas. In Chapter 4, Sue Atkinson describes her use of the 'plan, do, review' framework and in the next chapter, Maulfry Hayton explores more of these ideas, linking them to developing children's thinking through discussion. Following on from this, Rosemary Stickland shows in Chapter 6 how she uses her strengths and skills in language teaching to encourage questioning and scientific thinking in her children.

In **section C** there is a wide variety of stories, arranged roughly by age, starting with the youngest. We believe that the power of story is such that when you read them you will be able to enter into the experience with the teacher and children. The stories demonstrate a number of crucial aspects of teaching science: the importance of the teacher's role and the need for the teacher to construct a safe and stimulating environment where enquiry, questioning, exploring and all the many other aspects of being a scientist can take place and develop.

There is the need to provide children with the opportunity to share their 'intuitive' scientific ideas so that we can see how each child has come to understand his or her world. For example, in Chapter 7, Careen Leslie describes a unit of work in which she expands her children's knowledge about their bodies. She explains how she drew around the children's bodies on a large sheet of paper and then asked each child to draw what they thought was inside their body. The results were fascinating.

In other chapters we explore the idea that, in other areas of the curriculum, we have come to expect children to have 'intuitive' ideas. For example, in writing, the child is seen as an emergent writer, not just as someone who needs to learn to write. In mathematics, the child is seen as a mathematician in her own right. (This is explored in greater detail in Atkinson, 1992.) So it is in science. If we recognise the power of children's own ways of thinking then we are able to understand when children respond in unanticipated ways, able to plan more appropriately for individuals as well as the whole group, and we begin to appreciate that children often know a great deal more than we think.

This means encouraging children to communicate science in a wide variety of ways. Some stories show how teachers use concept maps to explore children's thinking both at the outset of a topic and then again at the end, so that they can assess what the children have learned. Other teachers use children's own books of various kinds, a science board with many kinds of recording, school assemblies and, of course, discussion to challenge and develop the children's understandings.

In **section D** there are chapters that pull together some of the themes from the stories in section C. As teachers of young children, we may not have been sufficiently confident to organise regular and ongoing scientific experiences for our children. It is not surprising since we may have been socially influenced to feel inadequate at science. The recent priority placed on science and the National Curriculum drive have put pressure upon us to reconsider this area as part of our teaching and to consider how we can cover the curriculum in a way that is meaningful to the children.

There are many areas that we could re-think and make different from our own experiences of the teaching and learning in science. In this book you will find that the teachers have focused on:

- greater enjoyment for children and for the teachers;
- greater enthusiasm for teaching and for learning science;
- greater participation by children – through doing and talking science;
- more opportunities to enquire – for children to ask their own scientific questions;
- more collaboration;
- more opportunities for girls to participate;
- initiatives taken by children for the planning of their investigations.

As adults, when we think of the word 'science', it often conjures up negative images for us. Unfortunately many of our school-based experiences may not have been positive. Often, we remember the science laboratory, the experiments that we had to conduct following a recipe from a book, card or chalkboard and often not getting the results we were expected to obtain – hence feeling a sense of failure. In addition, these scientific activities were sometimes totally divorced from our everyday lives. As a result, they held no meaning or relevance.

This is not the excitement that scientists sometimes experience. They have real problems to solve, they do not follow a recipe and they do not know the results before they start!

It may seem as if the world of science at school and the world of science for scientists are indeed far apart. Yet many everyday experiences are full of science. We engage in a range of activities that require a great deal of scientific knowledge and skill, such as cooking, yet we may never associate them with science. As adults we may even think that we do not have a great deal of scientific knowledge. However, we act scientifically each day when we control the air temperature in our homes and we use our knowledge about insulation, heat and convection. Research by Hardy *et al.* (1990) has shown that these perceptions tend to be held by many women and they have a poor self-image when it comes to science. Yet women engage in many scientific pursuits each day, but call them something else! The masculine image of science as Alison Kelly (1987) has shown, does little to help dispel this myth – that this negative image has arisen through the:

- numerical dominance of boys in secondary science classes (although with a National Curriculum this may now only be true for science beyond age sixteen as science is compulsory before that);
- majority of teachers in physics and chemistry being male;
- predominance of males in science text books;
- limited references to female scientists;
- focus on topics that are of interest to males.

Whilst our school-based experiences may strongly influence a view that science is maybe irrelevant and 'male', our home experiences may or may not reinforce this. In Chapter 2, Maulfry Hayton gives an account of her early childhood experiences in science. As you read this chapter you will be struck by the wealth of scientific experience she has had in her home and surrounding environment. Yet, as a child, she would not have viewed these experiences as science. The term 'science' tends to be associated with school-based activities – because we label these discrete, often decontextualised, activities as science. The relevant and meaningful everyday experiences that we have that clearly *are* scientific are not given this label – certainly not by most parents or teachers!

It is our hope that after reading this book when you hear the word 'science' you will associate science with reason, for both you and your children. We hope that your own concept of science and what it means to work scientifically in a variety of contexts will develop as you read. We also hope that as you try out the ideas in the book with your own group of children you will, over time, find that you teach science as confidently and as expertly as you teach the other curriculum areas.

Section A The challenge of teaching and learning in science

What does science mean for us?

1 — Approaches to teaching and learning in science
Marilyn Fleer

In this chapter, Marilyn explores four different possible approaches to teaching science. We probably all use different approaches at different times and for different purposes, but some clarity about these different approaches might help us to reflect on and improve our own practice.

> THEMES: background theory; discovery approach; transmission approach; process approach; interactive approach

Introduction

By trying to develop an awareness of the ways in which we teach science and the activity children undertake in the teaching-learning process, we can reflect on the purpose of what we actually do in the classroom. This type of reflection can help us improve the learning situation for the children we teach. You might find it helpful to reflect on these types of questions.

In our science lesson, were the children:

- left to discover things for themselves?
- told the 'right answers'?
- focusing on skills or content or both?
- asked what they understood and encouraged to ask questions?
- working together with the teacher on the science challenges?

The type of response to each of these questions will provide some indication of the approach taken by the teacher. For example, if the children were left to discover the answers for themselves, it is likely that the teacher was following a *discovery approach*.

If the children were being told the answers to most of the problems, with little involvement in the process, it was likely that the teacher was following a *transmission approach*.

If the focus of the lesson or experiences was predominantly on skills, such as observing, communicating and classifying, then it was likely that a *process approach* was being adopted. However, this approach is difficult to identify unless all the science lessons are observed over a period of time, since most scientific activity involves the use and development of process skills.

If the teacher actively tried to find out what children already understood and then set about encouraging the children to ask scientific questions, then an *interactive approach* was being used.

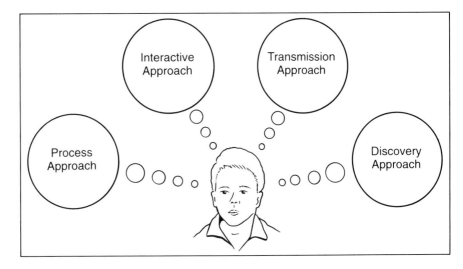

Figure 1.1 Which approach to teaching science should I adopt?

The reality is that most teachers adopt a mixture of approaches (see Figure 1.2). None of the approaches mentioned so far is ever implemented in isolation. Nobody will interpret the description of these approaches in exactly the same way.

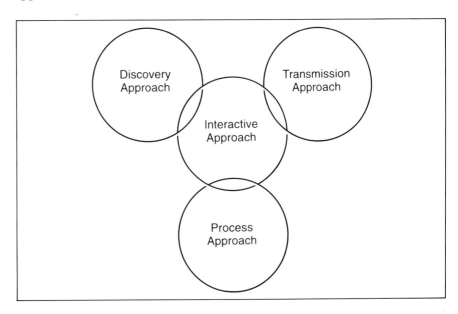

Figure 1.2 Four approaches to teaching science (Biddulph and Osborne, 1984: 5)

3

Individual teachers have their own set of beliefs about how children think and learn. Their personal teaching philosophy will influence which parts of an approach they accept or reject. The biggest challenge we face is trying out the different approaches, for unless we try different techniques, we will never be in a position to decide upon the merits or constraints of each. Acquiring a range of techniques and an understanding of when and how each can be used strengthens our teaching ability, since children respond in a variety of ways to particular approaches and will gain the most from their school experience by being exposed to different styles.

The discovery approach

A lot of curriculum support material produced over the last fifteen years has been developed with this approach in mind.

The role of the teacher	The role of the child	How science is viewed	How learning is viewed
• to select a range of materials and tools that will lead children to enquire about a particular phenomenon, for example, the conductivity of metal • to organise a sequence of carefully selected experiences that will lead children to see the patterns or characteristics across each activity, for example, plastics will not conduct electricity • to observe and physically assist children as they interact with the materials	• to use the materials and tools available • to see patterns and draw conclusions from the hands-on experiences provided	• as being out there to be discovered by the children, with the materials or activity setting the context for such discoveries	• as a process whereby children learn through direct observation and manipulation of the physical environment • as a process where children's development sets the context for their learning – when children are intellectually ready, they are more likely to see patterns and draw conclusions

Figure 1.3 The discovery approach

The advantages of this approach are:

- It is well understood that children will learn more if they are physically engaged, and this approach focuses on actively manipulating materials.

- Teachers of very young children have been using this approach successfully for quite some time now. The notion of a science table, a water trolley, a sand trolley, and table activities full of science materials and tools, such as magnifying glasses, seeds and magnets, is well understood and these are common features in many schools.

- Teachers feel comfortable with this approach, as it stimulates child interest and provides opportunities for the teacher to organise follow-up activities (perhaps through adopting a different approach).

However, the main disadvantage of this approach is that a teacher can never really be sure what children are learning just by manipulating their physical environment. Children cannot be expected to rediscover the major scientific achievements of our culture.

In addition, this approach does not take into account what children already know. If the teacher does not set up a situation in which children actively outline what they already understand, the materials provided may not be at all challenging to the children. They may only be reaffirming what they already know and understand. It may fail to focus children's attention on the relevant features of their experience and challenge their existing ways of thinking.

The transmission approach

The transmission model is an approach to learning that most people are familiar with. It was what many of us experienced at school.

The role of the teacher	The role of the child	How science is viewed	How learning is viewed
• to act as the central element in children's learning – the teacher controls the content, direction and pace of learning • to hold all the knowledge and to share this with the children, either verbally or through demonstration	• to be a passive participant in the teaching-learning process • to listen and (usually) record the information being shared by the teacher on paper for later learning	• as a body of knowledge to be conveyed to children	• as a passive process • as a process in which content knowledge is important, but skills and attitudes are not emphasised

Figure 1.4 The transmission approach

The advantages of this approach are:

- Information can be easily and effectively shared through this approach. In many instances, a teacher needs to outline a series of ideas, information or demonstrate some procedure to a whole group. This can be done quickly with large numbers of children using the transmission approach.

- Teachers are familiar with this approach. Children come to expect it at some stage in the teaching-learning process.

- In some instances, such as during care and safety procedures, this approach is the only safe method of sharing information with children (for example, when learning about dangerous chemicals).

As with the discovery approach, the main disadvantage of transmission teaching is that a teacher can never really be sure what children are learning. With the focus on the teacher, there is very little opportunity for children to express their understanding of the topic. As a result, it is difficult to know what sense children are making of what the teacher is saying or demonstrating. It is only when the transmission approach is used in conjunction with other techniques that a teacher has any sense of how children are interpreting the information.

The process approach

The process approach to learning has gained a great deal of popularity over the last ten to fifteen years. Many curriculum documents available to teachers, including national documents, have been built upon a view of teaching and learning that is skills based.

The role of the teacher	The role of the child	How science if viewed	How learning is viewed
• to examine curricula materials and identify the necessary scientific skills which will be focused on e.g.: observing, communicating, classifying, predicting, hypothesising and organising a fair test • to organise the teaching materials and resources needed for developing children's scientific skills, for example, by providing magnifying glasses, focusing questions and organising an excursion for children to develop their observation skills • to plan for a balanced curriculum for the development of all the scientific skills	• to participate actively in the learning experiences provided by the teacher • to develop all the scientific skills: observing, communicating, classifying, predicting, hypothesising and organising a fair test	• through the scientific methods of observing, communicating, classifying, predicting, hypothesising and organising a fair test	• as the active participation of the child in skill development

Figure 1.5 The process approach

The advantages of this approach are:

- Children are actively involved in learning.
- Learning sequences can be easily organised, since the focus is on developing particular skills and content is less significant.
- Teachers are familiar with it, since many curriculum materials and teaching resources are based on this approach.

The main disadvantage of this process approach is the concern for the limited attention being directed towards science content. Without some consideration for a balance between content and process, valuable areas of learning may be neglected. Similarly, a balance between the different learning areas in science is needed (such as life and living processes, materials and their use).

The interactive approach

Biddulph and Osborne (1984) describe this approach as including an element of each of the above approaches within a framework that begins with finding out what children know, encouraging them to ask scientific questions and then assisting them to answer their questions through investigations. This approach is less known, although many good teachers already use some of the features.

The interactive approach was developed by a group of researchers at the University of Waikato in New Zealand (Biddulph and Osborne, 1984), after extensive research into children's thinking in science (Osborne and Freyberg, 1985). These researchers were concerned with the scientific views held by children in both primary and secondary school. In their research, they found that many of those views were not those commonly accepted by the scientific community. In addition, they discovered that, even after traditional science teaching, most children's views did not change.

Figure 1.6 The interactive approach

The role of the teacher	The role of the child	How science is viewed	How learning in viewed
• to set the framework for children's learning and to organise an exploratory period in which children freely manipulate materials (designed to stimulate the children into thinking and questioning) • to co-ordinate learning by facilitating question-asking; providing materials and ideas for investigations; assisting the children with report writing; authorising the children to invite an 'expert' in the field to the classroom to help answer challenging scientific questions • to follow the children's interests and set up learning pathways	• to express his or her ideas about the topic under investigation • to ask scientific questions that can be investigated • to carry out investigations to answer his or her questions • to record the investigation and the results and share these with the class	• as a human construction which can only be understood within the context of cultural, social and historical considerations • as knowledge which is likely to be modified over time as human needs and ideas change • as a discipline where there is no one right answer, but many solutions to problems and needs	• as a human construction – children try to make sense of their world through active exploration of the environment and social interchange with people around them

There are many advantages to using this approach.

- Children are highly motivated when it comes to conducting their investigations, since they are answering a question that they are personally very interested in, and which is, in most cases, relevant to their lives.

- Children are likely to ask questions that the teacher would not have thought of posing (perhaps more sophisticated than the teacher would have imagined the children capable of thinking about or vice versa). Hence, science learning is more closely matched to the children's real learning needs.

- In most cases, the children are so motivated that they will find the resources they need to conduct the investigations, often involving their families and bringing in the resources from home. This reduces the organisational demands on the teacher and makes the children more responsible for their own learning.

- Learning experiences are broader and more focused. Topics are covered in greater depth, since the children all ask different scientific questions. Similarly, children spend longer on science learning, since their questions often involve a great deal of research. As a result, science learning takes a more integrated approach, as they measure and analyse their data and record and display their findings.

The disadvantage of an interactive approach is that it is difficult to introduce into more traditionally oriented classrooms and schools. Convention dictates that science content knowledge is not ambiguous, that there is only one answer. As a result, the acceptance of children's views as being equally valid to that of the teacher's creates tension for the teacher using this approach. Similarly, teachers using this approach are likely to be working with more than one type of science investigation at one time. If children are to be encouraged to ask scientific questions and conduct investigations to answer them, then a high level of managerial skills is required by the teacher. Obviously, a teacher new to this area would perhaps, through consensus, select only one or two questions to answer initially, until he or she felt comfortable with this approach.

Another difficulty with this approach is that children are not used to asking scientific questions, and the teacher must therefore actively work on this skill in the early stages. The younger the child, the more difficult it is for children to ask scientific questions that are investigable.

This chapter has given a brief view of four different ways of approaching the teaching and learning of science. Each makes different demands on the teacher and learner. Classroom life is generally much more complex than four clear approaches to science, as the stories in section C show.

We hope this book will help the reader to answer the question 'How do you approach science?'

— 2 — Science in early childhood
Maulfry Hayton

So where do children start from? Maulfry Hayton's chapter gives us a graphic description of how rich and fascinating a child's early experiences can be. The biographies of famous scientists often contain reference to childhood experiences like those described here. If we recognise this richness and build on that experience, our science teaching may benefit considerably.

> THEMES: the richness of early learning and current views about the ways children learn

At home

When I was six years old, I had my own 'museum' of treasures that I kept in a box beneath my bed. I added finds of personal interest: a broken stone that revealed 'crystals' inside, interesting shells, a small fossil and so on. We once found a dead Redwing: having identified it, with a bit of help, in the bird book, I later added the dead bird to my collection until my mother discovered it a day or two later – no doubt beginning to smell. For my brother and me, there was interest and excitement everywhere, often linked to a genuine desire to find out why?, how?, what if? Through our senses, we pursued endless explorations, gaining firsthand experience of the world and elements around us.

Water

Fascinated by the wriggling larvae of a mosquito, I remember wondering at the relationship between those water creatures and the air-borne insects they would become. At an abandoned market garden nearby, there were deep tanks of water with an abundance of water life; I spent hours gazing into their black depths, my *Observer's Pocket Book of Pond Life* in my hand. Water fascinated us: there was a stream in a wood to dam and divert, observing the resulting ox-bows. At the nearby beach, rock pools offered other, minute worlds. If we couldn't take home tiny crabs or pink, waving sea anemones, there was always a choice of seaweed; bladderwrack to pop, unnamed varieties to identify and some to help foretell the weather.

Air

I had been given a little, old, china donkey; it had a hole – for seaweed – where the tail should have been. Printed in gold letters behind the donkey was a weather forecast, relating to the condition of the seaweed. Sadly, frequent handling in the past had worn off some of the words: the last sentence read ominously, 'If . . . earthquake'! There were more promising ways of forecasting the weather; my mother kept a daily weather record in her diary and would triumphantly refer to previous years' records if someone was foolish enough to remark, 'Very wet for the time of year'. We learnt how to use a minimum/maximum thermometer, and my brother made a rain gauge that he set in a hole in the lawn, taking daily readings.

My mother showed us her old school 'copy book' in which, when she was eight, she had recorded that she had seen Halley's comet; she told us that, as an adult, she had seen the aurora borealis from northern Scotland. We were able to enjoy rainbows when the weather was changeable and in the spray from the garden hose. Indoors, an old, cut-glass door knob refracted the sunlight, creating miniature, indoor rainbows. We were introduced to the wonders of the night sky, adding to our understanding with the aid of grandfather's rotating star chart. Space travel belonged in children's comics and the future.

My brother's most adventurous scientific exploit was to put a bucket over his head, then go under the water in the rain barrel, breathing the air in the bucket.

Earth

We explored different habitats. With friends, we made dens and secret camps in undergrowth or trees. My brother dug an 'underground' camp – a large hole in the ground. Observing the subsidence after a day of rain, he sank an empty water barrel in it to shore up the sides – perhaps this was the origin of his life-long interest in geology and caves.

An open-air geological museum introduced us to specimens that were more exotic than those we could find ourselves, serpentine, richly veined marbles and fossilised wood. On a visit to the Isle of Wight, our pleasure at filling a glass lighthouse with different coloured sands was heightened by the knowledge that sand could be anything other than the shade on our local beach. Lump chalk found on an outing provided a tool for hop-scotch, and clay in the bed of a moorland stream provided material for experiments at making our own pots.

Fire

We made bonfires: feeble, smouldering heaps where we experimented with firing our clay pots, comparing those made from stream and garden-found clays. I was disappointed to find that the heat from the fire was not enough to melt broken glass to decorate my irregular dishes. Other pots we fired in the oven indoors, alongside the Sunday dinner, but the greatest pleasure came from the bonfire firing, blackened and crumbly though the pots became.

We 'cooked' poultry grain in tin cans, throwing the inedible contents on the garden; much later there was an interesting patch of assorted growth where our discarded experiments had taken root. We cooked potatoes, three-quarters charcoal but pronounced delicious. Burnt sticks in the fire provided charcoal to draw maps and write messages.

Other explorations

As young human beings, we were also inquisitive about other living things.

Flora . . .

For children, the half-acre garden offered a wealth of opportunities. We each had our own garden plot, though I suspect mine grew predominantly weeds. We could name many plants and trees, some of which offered edible snacks, whilst others were curiosities, such as Californian Poppies with their protective flower caps and a cactus plant that flowered only once in seven years. In summer, I made stinking 'perfume' with rose petals and water and pressed pansies and ferns; in autumn, I collected skeleton leaves and searched for fungi with strange names. One corner of the garden was plagued by common horsetail, fascinating because of its association with the time of the dinosaurs.

I ventured beyond the garden with friends, becoming familiar with other habitats. On a large tract of moorland, bog-cotton, sundew and purple marsh orchids grew. We listened to the curious sound of stonechats and knew where an owl lived in a hollow tree. A friend's grandmother had a pet jackdaw that talked, but I felt much happier watching wild birds.

My mother was a keen naturalist, and walks were always a rich source of discovery. If the name of a plant was not known, we searched for it in books on our return; at other times, we would search for details of a bird, tree, caterpillar or discarded egg shell. Walks had a purpose: we collected wild mushrooms, kindling wood and fir cones for the fire, dandelions and shepherd's-purse for the rabbits and soil from the molehills for potting plants. Finding blackberries was always worthwhile; the fruits that were not immediately eaten were used in the kitchen, some for home-made wine, with fresh yeast fermenting on a piece of floating toast in the large, yellow mixing bowl.

. . . and Fauna

Occasionally, I discovered nests of tiny, pink baby mice in the corn bin or listened to a hedgehog making a meal of stag-beetles at dusk. On sunny days, there might be slow-worms to be found in the pungent compost heap or tiny lizards on the rockery. In addition to the wild animals nearby, we had pet dogs and cats. We learnt to care for animals' differing needs and understood life cycles long before we recognised the term. We observed pet rabbits mating, watched the young litters develop, came to terms with the occasional cannibalistic doe and with death. There were ducks and chickens to be protected from predator foxes, and the thrill of finding a double yolk in an especially large egg. I knew that the hens required grit, so that they wouldn't lay curious, soft-shelled eggs, and that they would be alarmed if someone they usually recognised wore a very different hat – an early lesson in animal psychology! Temporary homes were often given to injured fledgling birds; they rarely survived the shock, and little graves were duly dug and decorated. We learnt to understand the physical needs of living things, and our responsibility towards them grew. At the same time, we found great pleasure in being close to so many small animals and gaining greater understanding of the natural world.

Winter experiments

Though summer provided the greatest explorations, winter held other possibilities. Making ginger beer was a great favourite for several years, with the added fun of the corks 'exploding' and a kitchen covered in sticky liquid. 'Eiffel Tower' crystals made another sweet drink when dissolved, and we made a drink called sherbet, from an old family recipe. Experimenting with cooking was always popular, though the ingredients must have been quite mundane: 'mixtures' were invented to be spread on toast and assorted fillings on bread, named 'surprise' sandwiches.

On cold mornings, I remember being puzzled by the similarity of fern-like patterns on ice covering the windows and the fossil-ferns in coal; I hadn't heard about fractals then. Steam rose from clothes aired in front of the fire, and we made shadow puppets with our hands in the light of our torches, when we were supposed to be asleep.

Specialist scientific apparatus was rare; I was once given a pair of toy, plastic binoculars on holiday in Cornwall, and another time found a small magnet in a Christmas stocking. Books offered to fill the gaps in our knowledge; books of our own, older family treasures and those added on regular visits to the library. A new encyclopaedia introduced us to the latest technology, including a photo of a lady in a ball gown made of glass fibre – we found the concept of clothes made of glass truly astonishing! Gloomy afternoons during any season of the year could be spent in the local museum exploring the exotic range of exhibits gleaned from home territory to further afield.

At school

The enjoyment of first-hand experiences out of school was not matched by anything we did in school. Nature study was on the timetable in the infant school, but I can only remember one walk in a nearby wood, when I collected some conker shells. At the junior school, there were beans growing in jam jars with pink blotting paper, though I didn't know why. Mr Hillier let me bring a pet rabbit to school for the day, but was angry when he found a wet patch beneath its basket. When I was nine or ten, Miss Galloway encouraged me to bring in butter and some wooden butter pats to make butter curls. It was disastrous; the weather was hot, and there was no refrigerator at school.

When I moved on to the next phase of education, a friend confided to me that we were to start 'science'. This was new as a subject, and I wondered what we'd learn; she replied ominously, 'cutting up worms'. I lived in dread of the day we began science, but worms were not on the syllabus. During three years of studying general science, I recall only the bunsen burners and something to do with litmus paper. Science was now a special type of learning: we had to write up experiments according to a format set by the teacher, and nothing we did seemed to me to relate to the real world. Even the room, with its special name 'laboratory', looked different, with its long benches and high stools.

Parents and teachers: working together and the role of adults

As an adult, I now recognise the rich scientific experiences I enjoyed, even though at eleven, I was unsure of the word 'science'. I don't think I had an exceptional childhood, but I did have a young child's natural and insatiable curiosity plus rich opportunities to explore things which were personally relevant and freedom to choose and do many things. Society has changed since today's teachers were children, and those we teach may not enjoy the same freedoms. However, the details – a garden or interesting surroundings – are less important than attitudes. We must bear in mind that children do not develop alone, so that from their first day of life, they grow in the company of others. Learning needs to be seen as taking place in a social context, with parents as the child's first educators. Chris Athey's (1990) suggestion of developing a professional partnership based on collaboration between teachers and parents could contribute a great deal to understanding. She suggests that teachers include parents in their work, valuing the parents' first-hand and deep understanding of their children. The value of socially constructed learning can be further understood in the light of Vygotsky's (1978) concept of the learner achieving more, in the company of someone who understands more than the child at this moment. Parents, teachers and other adults can all make vital contributions to the child's construction of understanding.

The implications for science in the early years

Young children are naturally curious; they are natural scientists in their eager searches for answers to questions about the world. Although I did not become a scientist, I have kept a childlike curiosity in the world about me, enabling me to share in children's curiosity and to reflect on ways of creating rich environments for learning in school. John Dewey's words hold as important a message today as when they were written in 1916:

> 'No one has ever explained why children are so full of questions outside of the school ... and the conspicuous absence of display of curiosity about the subject-matter of the school lessons.'

Writing about whole language learning, Ken Goodman (1986) listed what he believes makes learning easy; I believe this holds true for *all* learning:

> It's easy when:
> it's real and natural
> it's part of a real event
> it's whole
> it has social value
> it's sensible
> it has purpose for the learner
> it's interesting
> the learner chooses to use it
> it's relevant it's accessible to
> the learner
> it belongs to the learner
> the learner has power to use
> it

Ken Goodman's comments certainly relate to my own experiences out of school, ensuring that my curiosity and enthusiasm became linked inextricably with a love of learning and desire to find out, which is the basis of scientific enquiry.

In what I explored as a child it is possible to see the foundations of many strands of science; of earth and environmental science, biology, chemistry, physics, zoology, astronomy, geology, hydrology and meteorology.

The process of scientific enquiry will be supported by a sense of ownership of the learning, fostering positive attitudes. The principles of early childhood education, set out by the Early Years Curriculum Group, carry implications that are of significance for classroom practice. Quality, first-hand learning experiences in school offer opportunities to establish firm foundations for the learning of science. For children whose early, pre-school experiences offer limited exploration, it will be even more important that their early school experiences are rich and meaningful.

Research on the development and growth of early understanding of speech, writing and mathematics is beginning to demonstrate how babies and very young children construct their understanding in these areas. In the future, detailed studies relating to the beginnings of development of *scientific* understanding in babies and very young children will contribute to teachers' awareness. Approaches in science, as in all learning, should begin with what children *can* do, arising from young children's natural curiosity, which, if nurtured, is a limitless disposition.

This chapter contrasts the excitement of home learning and the lack of engagement with school science. This is a common experience for many of us, and in the research of the authors, it is most often described by women. As teachers, perhaps we need to consider the ways in which we treat girls in the context of science.

3 — Science from a child's perspective
Jonathan Osborne

In this chapter, Jonathan Osborne sets out to show us how children view aspects of the world. If teaching is to be interactive, it is important for the teacher to understand the child's perspective – what they already know and how they explain things. This insight into the child, particularly how they use language, provides a starting point for building on new understanding. So this chapter shows how, by encouraging children to be explicit, carefully questioning children about their meaning and focusing children's observations, we can start the process of helping them to look at the world anew, that is as a scientist would. This chapter is a useful link to the theme of section B, which looks at teaching strategies.

| THEMES: *interpreting children's writing and talking about scientific ideas*

Introduction

What do young children understand and know about science? If many adults have difficulty with even basic science – a recent survey revealed that only thirty-four per cent of a sample of two thousand adults knew that the earth went around the sun once a year – are we not being over-optimistic to expect young children to understand science? The short and brief answer to this question is a resounding 'no'. The evidence from sixty years of research, including the recent SPACE (Science Process and Concept Exploration) project, with which I have been involved, is that children are capable of expressing ideas about the natural world from a very early age, and that, although they may not be correct, they represent credible efforts by the child to make sense of their experience of the world. Scientists themselves are attempting to do the same thing simply in a more rigorous and prescribed manner.

The fact that children's ideas may rarely correspond to the scientist's world view is not important, since the opportunity to discuss and investigate natural phenomena is an essential foundation on which children can start to construct a scientific understanding of the natural world. The limited exposure provided by five years of secondary education is simply not enough to assimilate the wide range of concepts that modern science embraces. Primary science is a natural means of engaging a child's curiosity in a process of enquiry about the natural world that can and should assist conceptual development. But what are the features that characterise a young child's understanding of science? In this chapter, I have attempted to illustrate some aspects of children's thinking in science and their formative influences by drawing on a sample of the material that we collected in the SPACE project.

Children's expression of scientific ideas

Firstly, children *can* express their thinking about events and phenomena through talking, writing and drawings. For instance, in our research, children, when asked how they were able to see a book, gave the following range and type of responses which reflect different features in their thinking.

As a starting point for many children, particularly younger children, the process of vision appears to be non-problematic, as their drawings and explanations provide no indication that it involves anything other than the simple act of looking. In their explanations, they give no information, other than a simple description (Figures 3.1 and 3.2).

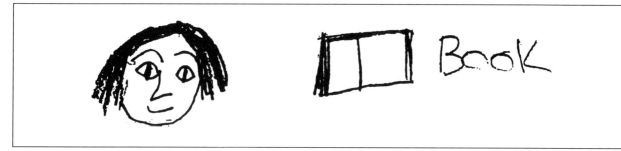

Figure 3.1 Age 11

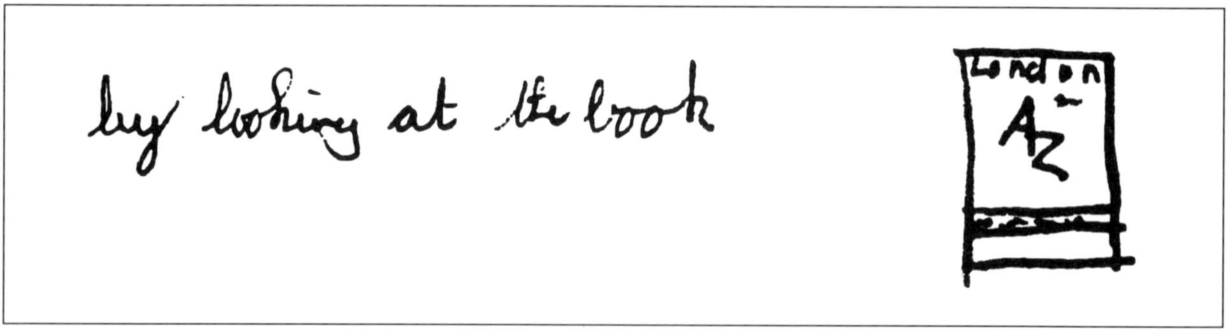

Figure 3.2 Age 9

Figure 3.3   Age 8

The next stage in a child's understanding is to give one single reason for the event, but without any explanation of the relationship. So, in Figure 3.3, the child ties the explanation of vision to the pupil of the eye, but fails to explain what its purpose is.

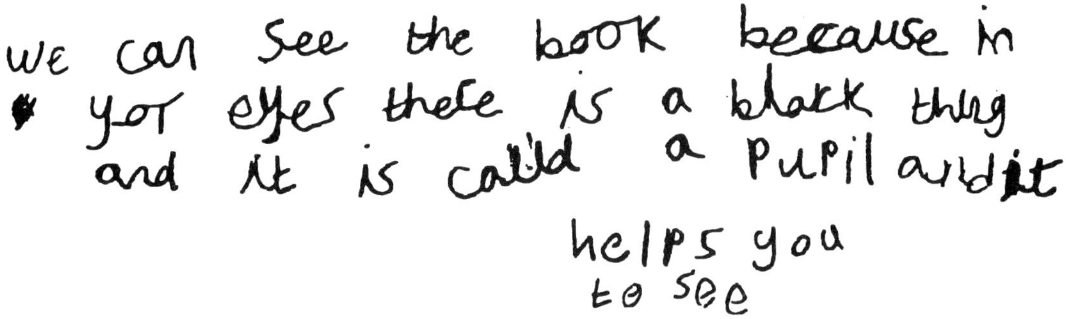

we can see the book because in yor eyes there is a black thing and it is calld a pupil and it helps you to see

Figure 3.4   Age 10

Figure 3.4 shows another example using a single reason, where the explanation is based on the idea that 'we need light to see with', but again does not elaborate what the role of the light might be.

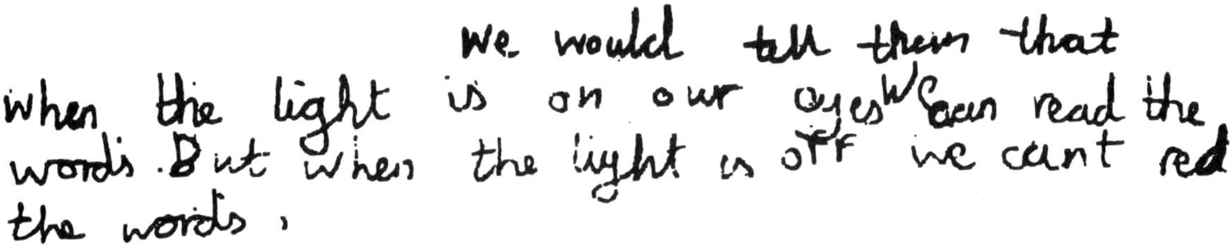

when the light is on our eyes we would tell them that can read the words. But when the light is off we cant red the words.

Many children also provide similar explanations in drawings that show a single link between the eye and the object (Figure 3.5).

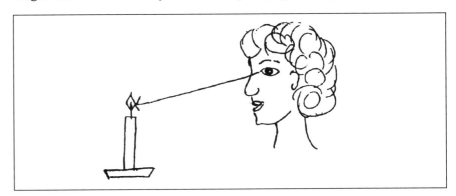

Figure 3.5   Age 10

Figure 3.5 is a fairly typical response and shows that the child sees vision as an 'active' process where an object is seen because of something coming from the child's eyes. Since looking at an object requires the action of either moving your head or your eyes, children extend this observation to imagine that the eyes direct some ray towards the object to make it visible.

A small but significant number of children recognise the need to show a link between the source and the object and the object and the eye to explain vision. They argue that 'light is necessary to see' and that 'we need our eyes to see with' and can show these two factors in a variety of forms. Figure 3.6 shows one such example where the child has attempted to reconcile these two ideas. Even if the idea is not scientifically correct, for the child it represents an advance, as it shows he or she is capable of offering an explanation *which links two ideas in a sequence*, rather than just a single idea.

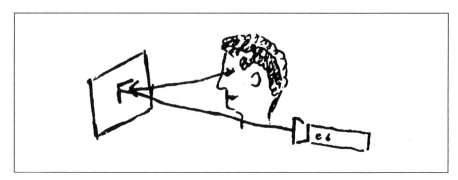

Figure 3.6   Age 10

It also shows the child trying to build in their existing ideas that vision is 'active'. This is a persistent concept, which leads to representations of vision which contradict a simple observation that the torch is emitting light. Figure 3.7 shows an attempt to reconcile this idea with the emission of light from the torch. Lines are drawn to the mirror and then onto the torch, but there is also a line from the torch towards the mirror. In the process of developing their thinking, children are almost certain to mix new ideas with older, intuitive ideas. As a teacher, it is simply best to recognise such ideas not as 'wrong' but as a worthy attempt to explain a natural event. Thoughtful responses which encourage the child to try to explain and then question their ideas by asking, for instance, 'how it is that, if something is coming out of our eyes, we cannot see in the dark?' are one way of posing a challenge to children's thinking which would aid the development of their understanding.

Figure 3.7   Age 9

Finally, there are a small number of children who are capable of understanding the full scientific explanation of vision, which is that we see because light from an object enters the eye (Figure 3.8). If a child has grasped the scientific explanation, it is important to recognise their success and praise them accordingly.

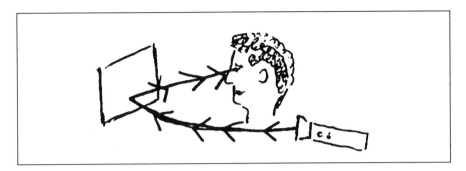

Figure 3.8 Age 11

What do children notice about the world?

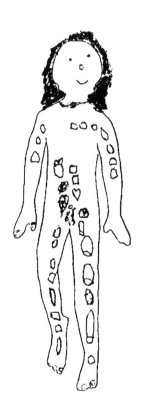

Figure 3.9 Age 5

Research has shown that the predominant aspects of events that children comment on are the obvious, external features. So when asked which part of the body is the most important, young children will list the external features, such as the eyes, nose and arms as being the most important. Only later do they become aware that life might be critically dependent on other parts of the body which they cannot see. Followers of Piaget would argue that this is because children are concrete thinkers who can always comprehend something more easily if it is a clear, tangible object which exists. Therefore the first organs children become aware of are the brain, through an awareness of their own capacity for thought, and the heart, because its beat can be felt. It is easy to check this view out for yourself by asking children to write down what is inside their body. On average, children around age five should produce lists with about three objects, whereas children of age ten or eleven will produce a much more extended list.

Another implication is that children's thinking is often limited to what they have experienced. In the following example, the children were asked to draw what happened to food inside their bodies. The response (Figure 3.9) shows that the child has no notion that food is transformed or broken down into other substances within the body. This is not surprising, as the scientific explanation requires an understanding that matter can be broken down into small bits and then altered irreversibly, an idea which requires the child to actively imagine matter as being made up of bits too small to see.

Children who are more advanced in their thinking do not show the food in the body, but will show a separate tube for the food and water to go down. Their reasoning is again based on the simple observation that, since waste food comes out as solids and liquids at separate points in the body, they must be processed separately within the body. Figure 3.10 shows much more sophisticated understanding, in that the child has a good, detailed knowledge of the whole of the digestive tract. This does not show whole pieces of food in the body and neither does it show separate tubes. The child's understanding is no longer based on simple observations, but on imagining what structure in the body would account for their observations – this child has moved forward in showing the capacity to apprehend objects which are only indirectly observable and those which are too small or too large to be seen.

What are the implications of this research for the teacher?

For the teacher, the implication is clearly that children starting to explore science should begin with making simple observations of what they can see around them. For instance, what kind of materials do we find in the world? What words do we use to describe them? What happens to the materials when we heat them or cool them? Simple questions like this help to introduce children to a range of phenomena which they may not have experienced and also introduce the language that is used in science to describe their properties. Thus the word 'boiling' has an everyday use to describe 'very hot', but in science, it has one very restricted meaning to describe the temperature at which bubbles of vapour are able to form in a liquid.

A further implication of children's concentration on external features and observable aspects is that their understanding and responses are often tied to particular contexts. The following examples from one child's explanations for vision show how responses can vary from one question to another within a remarkably short time without any recognition of the contradictions that this raises for an adult. In Figure 3.11a, the drawing shows the scientific idea of how we see and yet, in another item in the same interview, the child shows vision as being an 'active' process. The implication is that the child is unaware of the inconsistency and hence does not see any similarity between the two situations; that is, their answers are tied to the context.

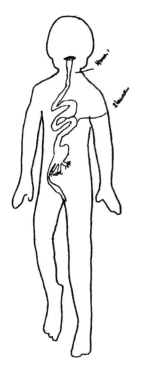

Figure 3.10   Age 9

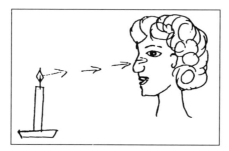

Figure 3.11a   Age 11

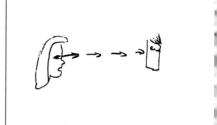

Figure 3.11b   Age 11

A second example comes from our work exploring children's understanding of simple astronomical events, where we asked children to show us how the sun moved during one day by adding the sun to a drawing showing some houses set against a horizon. We also showed them a picture of a tree casting a shadow early in the morning and asked them to show us how the shadow would look at midday. For the shadow cast by the tree, the consequence is that it will be shorter. However, our data shows that there was no link between children who gave the correct answer for one question and for the other, which suggests that the children see the two items as unrelated.

Thus, the role of the teacher is to help the child see what is common to the two situations and to concentrate their observations on the particular features that will help them to recognise both instances as similar. For the child who gave contradictory ideas about seeing, it would be sensible to point out that both cases involve the same process so why do their answers differ? A child who argues that it is impossible for the book to give off light should be encouraged to investigate whether it is possible for the book to reflect light. A child who cannot explain why a shadow is shorter at midday should be encouraged to chart the path of the sun across the sky and measure the length of a shadow at the same time. So by the careful choice of questions, teachers can point children to observe the particular features which are the basis of the scientist's explanation. And this is one of the important roles of a teacher of science – to help the child extract the relevant observations from the mass of everyday experiences.

Children talking about science

What are the features of children's talk about science? An examination of some of the statements reported for children in the SPACE work show the language being used descriptively to illustrate features and attributes of objects.

> 'Electricity is dangerous.'
> age 10

> 'When you push the switch, two wires connect to each other, and one of the wires goes to the bulb and the other goes to the cable.'
> age 10

> 'It (blood) goes through your veins.'
> age 6

However, there was another body of statements that showed children using language in a metaphorical sense, drawing on analogy to explain their meaning and possibly using it to create mental models to represent physical phenomena. The following are some examples.

> 'Electricity is like magic.'
> age 10

> 'Electricity is like gas ... you can't see it, it is dangerous and it helps things work.'
> age 8

> 'Electricity is like lightning that comes from space – it hits the wires that are on the street and it goes to the top of your house and makes the telephone work. All the electricity goes down the control box in your house.'
> age 8

> 'It (electricity) must go very fast ... faster than Concorde because you can phone to France in about ten seconds, so electricity can get to France that quickly.' age 10

The use of metaphor, analogy and simile

Similarly, in the research into children's understanding of the processes of life, instances were found of language being used metaphorically. Thus, the purpose of blood is to 'lubricate the joints' or 'keep your skin clean'. And, in attempting to explain how light travels and what happens to it when it hits a white card, the following child draws on an analogy with moving objects and their interactions with material objects.

> 'It pushes the air out of the way and then, when it gets on the card because the card is hard, the light can't get through, so it gets stuck, so you can see some light.'
> age 9

Using language in this way is very important in science, as it helps the child to construct representations of what the world might be like. Metaphors which compare the unfamiliar to the familiar are hooks which enable us to make mental pictures, which can then be used for thinking about objects which cannot be seen, e.g. blood, electricity, light. Ultimately, many of the theories of science are concerned with constructing pictures of things which are too small to be seen or too big to imagine; hence metaphor, analogy and simile are means which enable the development of scientific thinking. For the teacher, this means that they should be encouraging the children to talk and to describe what they have observed. The question 'what's it like?' is a useful way of developing such talk.

Using explicit language

More evidence of the difficulty caused in science teaching by the everyday use of language comes from the SPACE research on electricity, which showed that when asked to explain aspects of their understanding of this topic, children used the word 'electricity' almost universally in its generic sense with no specific explanation of its meaning. The following are typical examples:

> 'Electricity helps us in the home.' age 9
>
> 'Electricity gives us warmth.' age 9
>
> 'We use electricity for working things.' age 9
>
> 'Electricity is part of our lives.' age 10

The first three examples really refer to electricity as a source of energy. The final example shows that electricity is deemed to have properties which make it vital for life, and this was found to be very common. However, the everyday use of the word fails to differentiate between electrical charge, electrical energy and electrical current, resulting in confusion rather than clarity, implicit meaning rather than explicit meaning, and the teacher is faced with the task of encouraging explicit use of language through being explicit themselves and questioning children closely about what they mean.

Indirect evidence for the influence of concepts implicit in everyday language can be found in the research on children's understanding of light. Everyday language carries an implicit concept that vision is an active process – we 'look at' books, 'cast our gaze', 'stare daggers', have 'piercing stares thrown at us' and occasionally 'look right through' people. It is possibly not surprising, then, that the largest number of children think of vision as an interaction represented by a single link between the eye and object directed towards the object; that is, that vision is active as opposed to passive.

Common-sense reasoning in Science

Research undertaken in the past decade by Bliss et al. (1989), Carey (1985) and di Sessa (1988) has begun to explore how children's reasoning is influenced by common-sense reasoning. Joan Bliss and her co-workers suggest that children will often resort to two common explanations for why things do or don't move, which she calls 'support' and 'effort'. So a book is kept on the table by its 'support' and if there is no support, it will fall. This explains why we find it so odd that aeroplanes fly, as there is nothing visibly supporting them.

The use of such reasoning would also explain why infant children think that all you need to light a bulb is a single connection between the battery and bulb. Faced with a novel situation, they resort to common-sense reasoning, and the wire enables the battery, seen as the source of 'effort', to act on the bulb, the object.

Carey extends this argument to biology, arguing that biological processes, such as eating, having babies and growing, are seen by children as simply things which people do and require no explanation.

The problem for teachers of science is that much of scientific reasoning is fundamentally uncommon-sense, particularly scientific theories. For instance, common sense tells us that the world is flat, that the sun goes around the earth and that objects require a constant force to keep them going at steady speed. Science, however, tells us that the earth is a sphere, that it is the earth that spins and that, if there is no friction, objects keep going of their own accord. Although this only becomes a problem in primary science when teachers start to introduce the scientific explanations, the unnatural nature of the scientific explanation is a source of difficulty for children that needs to be borne in mind.

Conclusions

We see from this chapter that there are certain features of children's explanations and observations about scientific phenomena.

- Firstly, that even very young children are capable of expressing a range of thoughts and ideas about everyday experiences of the world through drawing, writing and talking.

- The explanations they give are often very context dependent, and they may not see the inconsistency between one instance and another.

- Children focus on the tangible and the concrete. Science gradually introduces children to objects which are too small and too large to see, and these are much harder for children to imagine.

- Children's explanations are initially descriptive and based on common sense. As they become more sophisticated, they develop arguments based on single reasons or, at a higher level, two reasons in sequence. They also begin to use metaphor and simile to construct analogies for objects which have to be imagined.

As teachers, we need to remember that the explanations of science are the products of some of the greatest minds of their generation, who have struggled to develop the ideas that we now so casually accept. For young children, engaging in science is the start of a similar, complex process – the difficulty of the path they follow should never be underestimated.

This fascinating chapter has given us some glimpses into the world of the child and has helped to set the scene for the rest of the book.

Section B Developing teaching strategies

How far do our teaching strategies encourage children to think for themselves?

— 4 — Children making paper planes
Sue Atkinson

This chapter shows one approach to developing and using investigations in the classroom. It is a simple but clear story which shows the learning and pleasure that can be gained from this approach to teaching science. The work is focused around a 'science board' and shows how it can be used to assist children to ask appropriate questions for their investigations. Hopefully, it shows that such an approach is easily repeated by the reader.

Age	7–9
Situation	teacher with a whole class; children working independently of the teacher using a 'plan, do and review' framework
Science	children exploring and planning their own investigations
Themes	children behaving as scientists; developing a consistent approach to learning across the curriculum; using a wall display to structure the science

Introduction

One of my big problems in the class is what to do with the other twenty-six children while I share books with a group of just a few. I want the reading time to be interruption-free, yet also want those children working independently to be working on something that is meaningful to them and that will cover a part of the curriculum. I also want to be as much a part of that work as possible, so that I can assess what is going on. The only way I find I can do this is to spend the first five to fifteen minutes of the session with children, planning their own work on paper (where possible or relevant), then sharing that in a circle so that I know what everyone is doing. Then I start the reading session with the few children, while the others carry out their plans, whether this is in science, art, maths or any other aspect of the curriculum.

I find that, when children plan their own work, they are motivated enough to get on independently. About fifteen minutes before the end of the session, I give a time warning, then there is clearing up, and we all meet together in the circle again for 'review time'. (Sometimes I make the review session about ten minutes, sometimes much longer.) Children show their work, and we discuss it and plan where this might lead us and what could be done the next day.

Of course, this is all quite time consuming to set up at the start of the year, and there is never enough time for each child to review their work every day, but over a week each child has a chance to discuss with the others what they have been exploring. In a one-hour session, it gives me about forty minutes for reading, and I find that, because the children have some measure of choice over what they are doing, they work well and with great enthusiasm. The work that they do in this independent way can be anything they choose that fits in with what is currently being worked on. This can include:

- construction with the Lego, Mottik, etc.;
- art and craft work;
- writing their story books and illustrating them;
- any ongoing maths or science that they want to develop;
- any new maths or science investigations;
- work on the computer;
- unfinished work from previous days;
- new ideas, such as 'can we make an aeroplane out of wood?'.

Planning a topic for the term

'Flight' was to be our topic for the term. The older children had been with me a year or so and had some experience of making their own books of science information (they were good at 'finding out' from books), planning their own investigations and devising fair tests and recording their results for review times. I hadn't yet had time to develop fair testing with the younger children, so that became one of my own aims for the work. I also wanted these younger children to be able to explore a wide variety of ways to record their work.

A lot of the children I had in my class I regarded as deprived, in that they had few books at home, so whatever I did, I had to leave plenty of my time for reading with individuals. Although the area around the school was an area of great poverty and unemployment, the children were mostly those who experienced gripping entertainment at home, from television and videos to computer games. Anything that I did at school had to be just as gripping. It had to be based on practical experience, and I was concerned to develop a consistency of approach in the learning in the classroom. I don't think that the children I teach separate their world into different bits of learning, and I was concerned to show coherence in the ways that I organised the learning.

The approach to reading was based on children becoming more confident and on their free choice of texts – I treated them as readers. In writing, they mostly chose what they wrote about, made their own books and discussed their work with me, individually or in groups, making finished drafts of work where necessary. I was led in this language work by what the children wanted to do and what they needed to do to extend their knowledge. Any end product (such as a book they had made being put into the class library) was very much a part of the child and, in a very real sense, was a part of their creative self. I wanted them to experience a similar thing in science. I wanted them to question what is going on in the world around them and to plan ways of answering those questions where they could, just as a scientist would.

I had experienced the older children becoming gripped by real science when we had done a project on bridges and buildings the previous year, but many of their investigations had fizzled out into nothing because there was a lack of time both to address all their problems and questions and to let them explore their interests in depth. I didn't want to abandon my classroom organisation of working in small groups, so I decided that:

- I would make much more of the 'plan, do and review' framework, so that everyone knew what everyone else was doing;
- I would give over one whole display board to science (this had worked well for maths);
- I would continue to encourage a wide variety of formats for recording work.

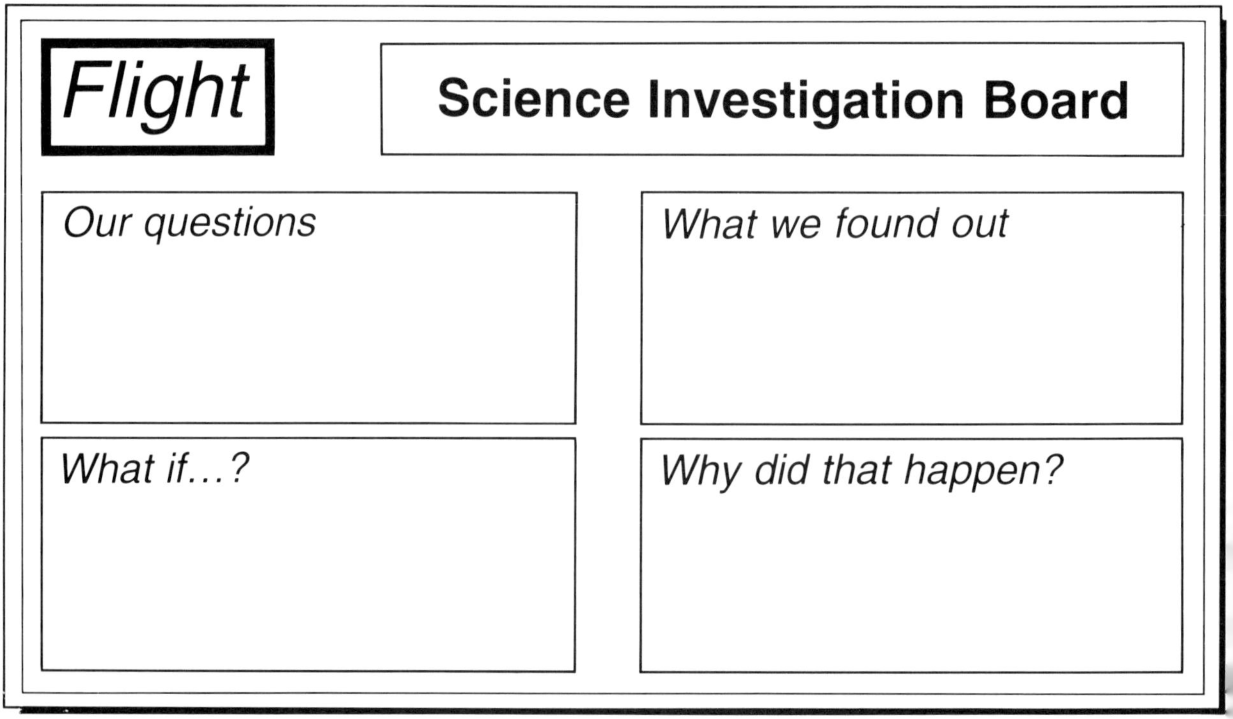

Figure 4.1   Science investigation board

Flight	Science Investigation Board
Our questions	What we found out
What if...?	Why did that happen?

Over a few weeks at the start of term, we developed a long list of the children's questions. (One way of doing that is illustrated in Chapter 6 by Rosemary Stickland.) Of course, many of the questions only arose once the children had actually made some planes and started investigating, but these first questions were important, because they revealed existing knowledge about flight. Some of their questions were:

- What is the furthest that we could make a paper plane go?
- Why do some planes work better than others?
- Which kind of paper is the best?
- Do some designs work better than others?

Many other questions arose during the term, particularly at review time.

Children recording their work

I had developed a great interest in the ways in which children recorded their work and in my ways of using these recordings to assess children's understandings. I encouraged them to put up any of their work on the science display board, and this was soon filled with assorted drawings, charts, measurements, examples of planes and statements, such as those following:

> 'My plane went 15 metres before it landed. I think it will hold the record.'   Kevin

> 'If you put a paper clip on the nose of this plane, it goes further. This one went 9 m 30 cms.'   Jason

> 'I can make my plane go to the right if I bend up the wing tip.'   Tracey

There were not many books available about paper planes, so the children made their own books, and these were read with great delight in the reading corner. We had one big class book that we read together at story time, but as children preferred to put their results on the board, this book was slow to get going. I therefore started writing things in it at review time, and the children liked that. I found it particularly useful to use simple zig-zag books with the younger ones, as these seem to encourage sequential thinking – first I did this, then this, then that happened.

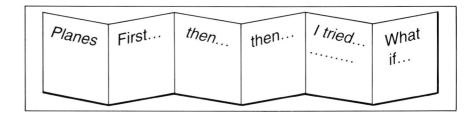

Figure 4.2 This kind of book can be added to as new ideas arise, or if the investigation needs to be re-thought.

One of the children's questions had been, 'how can things fly?' and one day, right at the end of a review session, this question was raised again. There wasn't time to answer it before lunch, so I put the question on the science board on brightly coloured paper and invited answers on pieces of paper that I put on the board inside a cardboard envelope. Children suggested their answers, and these included:

- the wings push the air down (one group had become fascinated by various investigations on a book on flight);
- things fly if they are light (I added to this, 'true for a bird but what about a jumbo jet?');
- things fly if they are the right shape.

I thought that last contribution was terrific and, in a sense, the whole term revolved around that question. Everyday things were put on the board, or things that were already on it were argued with and disputed and tested out. Parents and friends were invited in to see our questions and investigations, and the children encouraged them to contribute too. It was a genuinely interactive display!

Helicopters

I found that the younger ones couldn't cope with thinking about a fair test with the complexities of making planes – there were too many variables. So I did a session with this whole group (14 children) on making what they called 'helicopters'. This was entirely my idea and was initially teacher-led with the specific learning objective of clarifying for these seven year olds what they needed to think about in planning an investigation.

The basic design of a helicopter is shown in Figure 4.3.

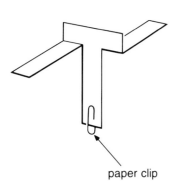

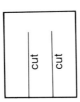

Figure 4.3 The helicopters are easy to make and to test.

I asked the children to explore what happened when they changed just one thing from the following list:

- the size;
- the type of paper (I had provided four different kinds to try to focus the thinking);
- the ways the flaps were bent;
- putting a weight, such as a paper clip, on the bottom;
- anything else they could think of – but just one thing at a time.

Before a new model was tested, I asked the child what they thought would happen. After they had stood on the table and let their helicopter go, I asked them what they had observed, why they thought that happened and what one thing could they change now?

This one session didn't in itself make them aware of fair testing, but in review time I asked all the older ones to say how they knew their test was fair when they discussed what they had done, so by the end of term, each child was quite used to being asked by me or another child, 'how do you know if it was a fair test?'.

Conclusion

The children enjoyed the work enormously. We gave an assembly on 'flight' that was so well received that we were asked to do it again for the middle school. As I thought at the time, its success was because I had followed the children's questions and interests.

I always find a difficulty between following the children's questions, needs and interests and covering all aspects of an overcrowded curriculum. My belief is that it is better to cover some things in depth and to give children time for quality learning, rather than to rush through everything.

I was reasonably happy with my 'plan, do, review' approach. I knew it worked well with younger children, and applying it to these seven- to nine-year-olds seemed obvious; it was a way of teaching that I found creatively satisfying and which also gave good results in terms of the children's learning. Even the children who find any kind of recording difficult were able to record with pictures and their 'developmental' writing.

This 'plan, do, review' approach blended well with the science board; further work with a science board is illustrated in Chapter 12. Children's recording of their science can be very varied, as can be seen in a number of chapters in this book; this is explored further in Chapter 18. Learning to do science is as important as learning about science. This chapter has shown one way of harnessing children's curiosity to do science investigations and begin to understand what it means to be a scientist.

5 — Talking it through: young children thinking science
Maulfry Hayton

This chapter is fascinating because of the powerful ways in which Maulfry channels the children's talk in class to develop effective thinking skills, which she then uses as a 'tool' for learning science. She was inspired by a BBC television programme on philosophy for children, which she followed up with some work with her children. She became involved with SAPERE (see References) and has continued to explore this work with her colleagues and children, developing 'communities of enquiry' for many subjects across the curriculum.

Age	Years 1 and 2 year 2 (first with four- and five-year-olds, then with six- and seven-year-olds)
Situation	groups with teacher
Science	free-ranging
Themes	using enquiry-based discussion groups in science; mind-maps

Can talking help children develop understanding?

A special kind of talk

Talking as an activity has now won approval in our classrooms. A wealth of research relating to the development of understanding has come to our attention, including that by Vygotsky on the relationship between talking and thinking. One study (Tizard and Hughes, 1984) compared the quality of talk at home and in pre-school settings and showed that much of the talk in homes was richer. The National Oracy Project (see Lalljee, in press) highlighted valuable areas that teachers could explore, and, with the inclusion of '*oracy*' in the National Curriculum English document, talking is seen as a specific area and worthy of inclusion. At last talking in class is valued, though of course some types of discussion will be more useful than others.

In 1990, *The Transformers*, a series of three educational television programmes, was broadcast. The series focused on certain aspects of education, challenging children to reach out for their potential.

The programme which especially fired my imagination was on Matthew Lipman, whose philosophical discussion groups, or *'communities of enquiry'*, help develop skills of logical enquiry. Although I did not have a background in philosophy, I was excited, and began to see the value of this kind of talk in my classroom. It appeared to capitalise on young children's innate curiosity; to develop both thinking and listening skills; to broaden language skills and to enhance self-esteem.

Practical issues

The reality for classroom teachers in most primary schools may not often allow for work without interruption with one group for long. In my class of thirty Reception and Year 1 children, we *have* had some rich and exhilarating whole-class discussions of a 'Lipman' nature. However, in order to involve a greater number of children in an active role, I prefer to work with a smaller group. Whatever the number of children, the approach is the same.

We all sit in a circle

I prefer to sit on the floor. For me, it is important that I sit at the children's level, so that I am as near an equal member of the group as possible. In a circle, each member of the group/class can see everyone else.

Simple 'rules' made by the members of the group can be helpful. In my class, the children suggested:

- only one person talking at a time;
- everyone can have a turn;
- share your ideas;
- look at the person you want to speak to and/or say their name.

'Rules' apply to the teacher and other adults present too.

A 'trigger' is needed to start the discussion

In order to have a focus for the discussion, something is needed to spark ideas. For a discussion with a scientific emphasis, we have found the following useful:

- a child's comment or question about a topic of personal interest: 'Why is it light when I go to bed?', 'What's clay made of?', 'What would happen if . . . ?';
- something which a child has brought in to show – a newly sloughed snake skin, a torch, a sprouting potato;
- interest arising from a story, e.g. 'Could a whale *really* live in a garden pond?' as in 'Dear Greenpeace' by Simon James, 'Why *do* people put rubbish in the sea?' prompted by John Burningham's 'Oi! Get Off Our Train!'.

I have found that almost any quality children's picture story book will offer some possibilities. It is not necessary for the school to purchase a special selection of story books for science, though some obviously highlight certain issues better than others. Other examples of triggers might include:

- a small group science investigation, such as making paper aeroplanes or melting chocolate;
- a toy, e.g. something that works by remote control, a balloon, a balancing toy.

Two aspects are of vital importance in order to promote the highest levels of logical thinking: *open questioning* (highlighted in bold in the dialogue) and *managing the talk*. I will focus on these aspects in the following dialogues, in which I explored Lipman's ideas further with my next class of seven year olds.

Discussion about electricity

The children had been doing some work on static electricity

Amy: Sometimes, when I've been asleep, I wake up, and my hair's all standing up, and mum says, that's, um, electricity in your hair, and so, sort of like when I, it seems like there really *is* electricity in your hair, and um, if you rub it, the side of the balloon goes all sticky. *(They had used balloons to explore static electricity.)*

Karen: When, 'cos I think, when you rub a balloon on Mrs Hayton's jumper, it's full of, um, sort of . . .

Simon: It's natural fibre!

Karen: Yeah, well, when you rub, 'cos, well, if you rubbed it on, if you rubbed it on Mrs Hayton's jumper, which is made of wool, we could probably bring static – and sheep's wool as well. *(Lara followed this up – see Figure 5.1.)*

Teacher: May I ask something? Is that the same sort of electricity we are talking about, that actually powers this tape recorder? **(Teacher encouraging clarification.)**

Chorus: No, no.

Simon: It's made by a different power.

Amy: Plug power! **(Note that as in developmental [emergent] writing and mathematics approaches, the children find it possible to construct their own terms. These approximations fit the children's current needs and are likely to be more meaningful than the 'correct' terms when used in 'free-range' discussions such as this.)**

Figure 5.1

a girl is rubing her hand on a scabul net She miyt get a rowp burn it will get hot like eletriste

Figure 5.2

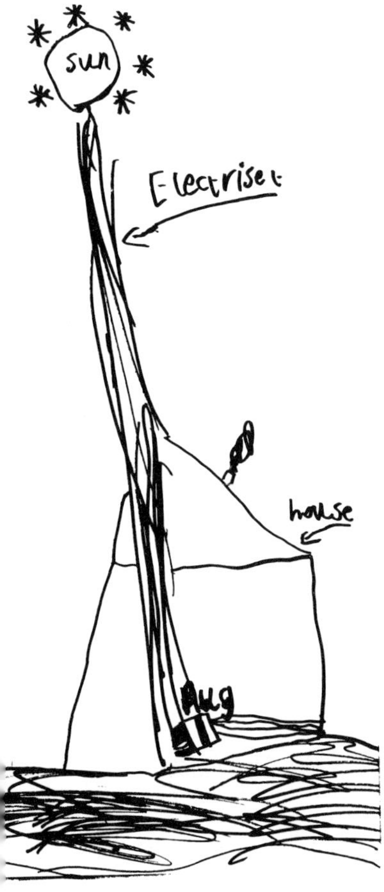

Teacher:	Can you tell us more about that? **(Teacher looking for meaning, defining terms.)**
Amy:	That's plug power. *(She points to electrical socket on the wall.)*
Simon:	Because, there's, um, a satellite up in space, and the sun powers down to that satellite, and then it goes down to earth, and they have big power stations, and the power goes into them, and there's lots of wires and stuff under the ground, and power can travel through them and go to houses and schools . . . and bungalows and places like that . . . So it will go through them and make things work like the tape recorder. *(Simon's drawing of electricity can be seen in Figure 5.2.)*
Teacher:	What *is* power? Has anyone got anything else they'd like to add to what Simon says? **(Teacher looking for clarification, encouraging reflection, giving time, inviting other opinions.)**
Lara:	Sun . . . hot, that goes down . . .
James:	If the cloud and thunder . . .
Amy:	Simon was talking about the sun's power and the plug power and different, um . . .
Teacher:	Can you tell us in what way it's different? **(Teacher, seeking explication.)**
Amy:	Well, it's different because it's a different *kind* of electricity, and because you can't rub anything on the sun, and you can't rub it on the power . . . To get it out of the plug, and there's a special sort of thing in the plug that makes all the electricity run through . . . But we're only talking about *little* pieces of electricity going through.

Lara: I think it's something to do with when you rub something or get something for a long time . . . It's like a rope burn, it gets very hot, and that might be how you get electricity!

Simon: Yes! I agree with you! Because if you rub your fingers on the floor, they get hot, hot . . .

Lara: Yes! Yes!

Amy: Um, when I fall over, on carpet, um, I fall over, and it comes down such a crash, it *really* hurts, and it burns my leg!

Teacher: Hang on – what's happening, with what you two said: rubbing your knee on a carpet and rubbing your hands and making a rope burn – what's happening here? **(Teacher, helping the children to see if there is consistency in what they are saying.)**

Amy: It's getting hotter and hotter . . .

Simon: Friction!

Teacher: Can you tell us something about that? Might that have something to do with electricity? **(Teacher, trying to help children clarify meaning and consider implications.)**
(This was one of those occasions when, as soon as I had spoken, I regretted it. I still have to work at biting my tongue.)

Simon: Well, the sun would probably go into the satellite, and then down to the earth, and then it would go past a machine or something . . . and it would make *friction* happen, and then it would go down into tubes underground, and then it would make more friction, and more sparks so . . .

Teacher: So . . . are you suggesting . . . does friction make electricity, or can it make heat? **(Teacher looking for meaning: what is the child suggesting?)**

Enabling questions

Without a doubt, it is the quality of the teacher's questioning skills, coupled with the way in which the discussion is managed, that makes a community of enquiry so special and contributes to the development of a wide range of skills. Questions need to be of an 'open' nature whenever possible. Matthew Lipman believes that the promotion of high-level thinking skills is best supported by questions which have their roots in philosophy. The teacher's questions in the dialogue here are of this nature, as are the following:

- 'You're saying that . . .' *(aiding expression and valuing child's contribution)*;

- 'Aren't you thinking that . . . ?' *(is the child making an assumption?)*;

- 'Why do you think . . . ?' *(what reason do you have?)*;
- 'Couldn't it be right that . . . ?' *(is it valid?)*;
- 'How do you know that . . . ?' 'How might we find out whether . . . ?' *(how could we get evidence to support what was just said?)*;
- 'What would happen if the opposite were true?' *(alternative possibility)*.

Further examples of enabling questions may be found in Matthew Lipman's books.

The dialogue continued with much interest and excitement. The children are always enthusiastic about linking ideas and making connections.

James: Well, I think thunder and lightning does make . . . Thunder and lightning is power as well, 'cos . . .

Simon: Light! It's probably friction ... up in the sky! There are two clouds that are quite hard – they hit together, and the, er, it makes friction. But then, as clouds are quite hard, they hit together, and then, er, it makes friction, but then, as clouds are much bigger, they make bigger friction. *(Afterwards, I thought that perhaps he meant that the thunder and lightning would be 'big', because big clouds would cause friction – 'bigger than a rope burn'. Some discussion on the properties of clouds followed: 'clouds are very light', 'clouds are not hard', 'they just go past each other'.)*

Amy: You've got a point. Well, it's a little bit like welding, because, when my brother and dad weld, lots of sparks come out, and that's because the thing was so hot, and it's rubbing against a cold thing, and making them both *really* hot!

Taking turns

Some children do not find it easy to take turns in a discussion. There will often be one child who, whilst having a great deal to contribute, is less sensitive to the needs of others. Because the discussion is a *shared* one, the group must share responsibility for its management too. In practice, one child will usually point out to the dominant member of the group that 'it's not fair', justifying their comment and making suggestions. Other group members may also contribute comments, whilst showing a surprising ability to tolerate. Such 'management' of class or group discussions can require a leap of faith for teachers, since we have often come to expect to be 'in charge'. Empowering others can require us to share our power so that our teaching ceases to be understood by the children as the process of them finding out what 'answer' it is we have in our mind. If we can move away from this guessing game and share power with the children, this helps the children to take responsibility for their behaviour and negotiate – both valuable skills for life.

Re-starting a community of enquiry

When taking part in a discussion like this, the children's enthusiasm is clearly evident. In their search for understanding, the children share and develop each other's ideas, and often maintain a lengthy dialogue. I find it is always possible to re-start the discussion on another occasion, if we have had to break for lunch or some other school routine. For example, the following questions can help elicit topics and further questions.

- 'What did you find interesting about . . . ?'
- 'I was really interested in . . .'s idea about . . . Does anyone else have anything they'd like to add?'
- 'Amy, you said: ". . .". What did you find interesting about that?'
- 'Yesterday . . . said that . . . . Is that always true?'

Keeping track of the discussion

Matthew Lipman suggests that teachers should make notes whilst the discussion is in progress, and many teachers who have used 'communities of enquiry' choose to note comments on a black/white board or flip chart. Because I prefer to sit down with the children as a member of their group, I found this impractical and uncomfortable. I believe it may also shift the emphasis back to *teacher*-centred talking rather than a community, with its emphasis on shared meaning. I do occasionally jot down some notes on a sheet for us all to share, and add names or initials by comments. Sometimes I tape the discussion so that the children can review it later; a taped discussion can itself act as a trigger to further talk. For the teacher, a taped 'community' is a valuable aid for assessment and future planning.

The discussion about clouds moved on

James: *(talking about the temperature of water in a swimming pool)* The top's hot, and the bottom's cold.

Amy: Yes, yes, but it might . . . But water isn't *solid* and won't just stay on top – it'll actually go down to the next layer . . .

Simon: And it'll soak it up . . .

James: Well, what I think happens is, if there is . . . *(They were off again. From here, the talk led to hot water and electricity, moving through batteries, energy, power – and 'running out'.)*

Amy: It runs out.

James: Like the light . . .

Simon: Like us, if we're tired, we can't think very much . . .

James: We run out! We run out of brain power! **(Here is another of the children's explanations, an alternative framework: 'brain-power'.)**
(A discussion on food for 'fuel' then followed.)

Karen: Like cars! Cars have to have more petrol and batteries.

Amy: Everything! We need to have food – like fuel, and batteries are like vitamins . . .

Another way of looking at things

Statements like 'plug power' and 'we run out of brain power' indicate that the children have their own, highly personal understanding of certain concepts. Through discussion, we are privileged to share this. The fact that their understanding is not in line with accepted scientific wisdom does not imply that the children are 'wrong', but rather that they have alternative ways of viewing the world. These *alternative frameworks* will be appropriate for the child's developmental stage, and tell us a great deal about what they *do* understand. At the same time, the teacher gains information about areas where individuals may benefit from support. I treasure the children's unique contributions and personal insights.

New knowledge

However, clearly being *told* that something is so is not the same as understanding. The child needs to construct that understanding for him or herself. This process will need time and a variety of future opportunities to explore concepts and make personal sense of them. An interesting topic towards the end of this discussion shone light on two children's current thinking of the way in which coal had been formed. One child was certain that it was created from 'molten lava'. At this point, I offered a simple explanation of the way in which coal had been formed millions of years ago, from rotted trees and plants. Amy tried to accommodate this new knowledge: she said, 'that's why sometimes you see coal mines, that's where all the trees fell down'. Therefore [she reasoned] coal mines are holes in the ground into which trees have fallen, rotted and turned into coal. In her book (Driver *et al.*, 1986), Rosalind Driver provides a number of interesting examples of children's understanding of scientific concepts, showing how children can find it difficult to take in new 'facts'. Maybe we can facilitate this process by discussing the idea of 'new knowledge' with the children and asking them when something is 'new knowledge' for them. Children find this exciting and openly discussing what is new can encourage discussion and enquiry.

Mind maps

Mind maps, sometimes also called brain-storms or concept maps, allow us to represent ideas visually. It is often suggested that these are made at the onset of topic, lesson or discussion, but this cannot allow for children's spontaneous and divergent ideas. Creating a mind map *following* a discussion can be a more valuable exercise for the group. The teacher's notes or a recording made during the discussion make this possible. An example of a mind map which traces the beginning of the first dialogue in this chapter is shown in Figure 5.3.

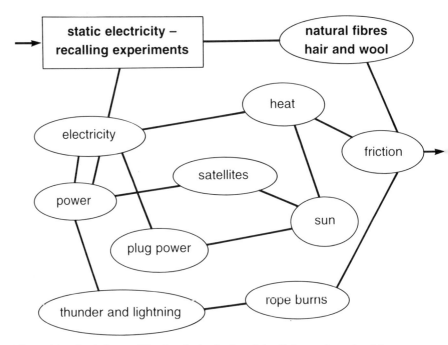

Figure 5.3   A mind map following the beginning of the dialogue about electricity.

It would be interesting to represent an entire discussion in this way, showing the links between both the ideas and the children's minds, requiring a huge, sophisticated three-dimensional model. Because space does not allow for this, a simpler linear illustration is shown in Figure 5.4 with the children's ideas given in the order in which they were introduced. This outcome was the product of only one, half-hour discussion. There are likely to be links between such 'free-range' discussions and mind maps, and the way in which the brain categorises, memorises, adapts, retrieves and builds knowledge about the world (Edelman, 1992). The children find it interesting to explore part of a discussion with a visual mind map (rather than a list). Seeing a visual representation of how one idea led to another, they often find they have further insights or choose to discuss one point in even greater detail.

Static electricity, natural fibres, hair, wool, static electricity, power, 'plug' power, solar power, satellites, power stations, 'source to outlet'*, clouds, thunder, rope burns, hair, static electricity, fall on carpet, friction, heat and electricity, thunder and lightning, friction, clouds, welding, sparks, cold and hot, stars, fire from the sun, clouds – not heavy, clouds are water, hot water, power of heat, top hot – bottom cold, water not solid, air temperature – warms water, sun has power, sun equals light and heat, sun equals electricity?, electricity, 'source to outlet'*, people control electricity, electric fence, insulation, need for circuit, safety with electricity, electric fence – jump, shiver, shiver go through arm, vibrations go through metal, 'electricity' and 'power' – same?, electricity is strong, different power needs, cars and factories – needs for power, power – length of time 'on', friction, light, cars – engine/motor, petrol, power inside battery, inside battery – another form of electricity, one car battery could have power equal to thirty small batteries, flat batteries – power runs out. We run out of brain power, energy, fuel, food – like fuel, need for food, batteries – like vitamins, food – changes to energy, goes round body, gives us power to keep us going, we can run out of energy, panting – need for more oxygen, dogs pant, power and energy, food – potatoes, water and sun, people need sun to live, trees help, water from taps, solar power, 'source to outlet'*, power stations, coal from molten lava, coal mines – trees fell into mines, trees need water and sun, to have everything we need the sun, nothing would grow without the sun.

*This was the only term that did not originate from the children. It has been used here for the purposes of economy – the explanation used by the children was too lengthy.

Figure 5.4 I was astounded! Seeing the connections made between ideas and the links they made in their own minds and between the minds of others in the group was very exciting. I had no idea that already by the age of six, the children had some understanding of aspects that touched on so many broad areas of science.

Carl Rogers (1983) believed that pupils need to be truly involved in the search for meaning: he saw the teacher's role as that of facilitator and this chapter has described one way of achieving this. Holding classroom 'communities of enquiry' may be one very positive way of facilitating learning and contributing to a process where meaning is negotiated.

This chapter shows how one teacher gave prominence to children talking to each other to explore their ideas. It shows how crucial the role of the teacher is during this type of discussion and how such experiences are essential to developing fluency with the language of science. The mind maps provide a valuable tool around which children can clarify their ideas and build on them through discussion. Further resources and suggestions to facilitate this type of discussion are given on page 176.

6 — Starting science from talking, listening and questioning
Rosemary Stickland

We have included this chapter because it is by a teacher who is – like many teachers – very much more confident with language teaching than with science. She shows how she builds on her own strengths and approaches science from her language work. She works with the children to search for their questions and approaches her science work from those questions.

Age	5–6
Situation	teacher with a whole class
Science	materials used for clothes
Theme	developing scientific ideas through talking, listening and questioning

From the very beginning, I started talking, listening and questioning with the children, getting them talking and listening to each other. (See my chapter in *Frameworks for talk*, Lalljee (in press).

Initiating talk and developing questions

I started this through having a 'news time' session on a Monday morning. We sat in a circle on the carpet, and I emphasised the importance of being able to see everyone in the class when you were talking. I encouraged them to look at each other while talking, and we played a few games to help them use eye contact. At first, the children were happy to talk about something that had happened to them at the weekend, still in the circle, then I got them to sit face to face with a partner and take turns to tell each other their news. I gave them a few minutes' practice and a lot of encouragement. Then I brought them back to sitting in a circle and asked them who could remember their partner's news and would like to share it with the class. This encouraged them to listen closely to their partner and take pleasure in someone else's news.

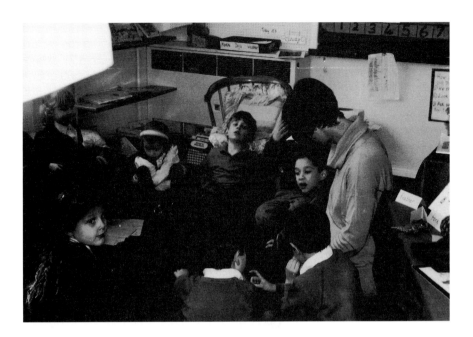

Figure 6.1 Sitting in a circle, discussing with teacher.

As well as learning to listen, children also have to learn to ask questions. When they first start school, they tend to bring in a great many objects, and having each child coming out to the front with their 'treasure' can get rather tedious. So I decided to change the format. I got the children to sit in a circle and we put all the objects in the middle. I asked if anyone had any questions about any of the objects. At first, some of them didn't know what a question was, but they soon learned from the others.

I then asked them if there was anything they wanted to find out about one of the objects, or if they were curious about anything. The first question was usually 'who brought that in?' We had a few weeks when we didn't progress beyond that kind of question. Then they started to say, 'where did you get it?'; that was an improvement, but there was only one answer to that, and I was hoping to move beyond that. I was trying to get the child who brought the object in to tell its story. It was a long, slow process getting there – as a teacher, you have to give examples of good, open questions and the children got used to me saying, 'why did you bring it in?' and 'what does it do?'.

The hot seat

Running concurrently with this work on questioning, I was putting a child in the 'hot seat' if I knew they had something interesting to talk about. One day, Stephen had been to the beach, and I was trying to get the children to see the difference between an open question and a closed question. Mostly the children asked closed questions, such as 'did you like it?', to which there is only a 'yes' or 'no' answer. I was trying to get them to think of the more open-ended questions such as 'what did you do on the beach?', but it took a while to achieve this. In general, children do see the difference between open and closed questions and gradually stop using only closed questions, but it takes time and you have to keep leading them and giving examples.

The 'hot seat' works because the children love taking part, and they make noticeable progress, both in the nature of their questioning and their confidence in their answers. For example, asking early in the year why a child brought something in will elicit the answer, 'because I like it'. Later in the year, the question will change to 'where did you get it and why did you bring it in?', and the answer is often sophisticated and complex. I find this way of developing questioning helps a class to bond, as they are genuinely interested in the answers and they really love the way the child in the 'hot seat' has control over who asks the questions.

When the new intake of children arrived after Christmas, it was clear that the others had made a great deal of progress. The new ones just kept making statements – one child, when we were in the middle of questions, said, 'I like my friends'. It was a statement that had absolutely nothing to do with our discussion. It was then that I realised how much progress the older ones had made the previous term.

Topic-related questions

Our topic for term 2 was 'materials', and it was very natural, with our background of questioning, to start them off with questions they had about materials. I was expecting them to ask questions to which I knew (or thought I knew) the answer, but they didn't.

We started off thinking about the clothes that we were wearing, and I asked them what they wanted to know about clothes. Instead of having them all asking questions at once, leaving me to select one child at a time to talk, I asked them to think about it in pairs. This is one way of ensuring that every child thinks about the topic and contributes something. I then collected the questions by getting them to sit in groups of four with clipboards. In this situation, they usually find it quite difficult to think of questions, so when I heard Helen say, 'what are T-shirts made of?' I stopped the whole class. I told them that was a really good question, and this generated other good questions.

I circulated around the groups, ready to help them when they got stuck. I then gathered in all the questions, and we put them in a 'floor book'. The idea is to make a big book to which the children can all contribute and which they can read in the book corner. It is generally very successful.

The questions in the floor book were:

- What are clothes made of? (Rosie)
- How does the wool get all joined up to make the clothes? (Annie)
- How do you make clothes? (Chloe)
- What are T-shirts made of? (Helen)
- What are blouses made of? (Stephanie)
- How do you get the Millbrook sign on our sweatshirts? (Elspeth)

It was interesting that two children didn't voice their questions at the time of the general discussion, but chose to wait until I was sharing a book with them on their own to ask a question. Perhaps they needed the time with me to 'ponder aloud'. But I showed at the time how valuable I thought the questions were by writing them down and following them up. Robert asked as we read together, 'if you mix glue with paint, would the paint stick on material better?' (He was thinking of the Millbrook logo on our sweatshirts.) I wrote Robert's question up on the wall and later put it in the floor book; I also gave him and a friend time to test it. They then displayed their results on the wall.

Then, over the next eight weeks, we set about answering these questions. The sorts of activities we did were:

- looking at various materials through magnifiers (this was a really good activity and they loved it);
- paper weaving;
- exploring dyes with fruit and vegetables boiled up with bits of white cotton cloth;
- printing on materials using different paints and paint mixed with glue;
- producing a display of different types of cloth, e.g. wool for a jumper, satin for a party dress and cotton for a T-shirt.

We would have carpet sessions (all the children sitting on the carpet together) to go over what we now knew, and the children made contributions to the floor book, including drawings, further questions and any answers that they found.

Figure 6.2 *It is crucial in science teaching that the learner engages in a personal way with their own questions and the quest for their own answers.*

Figure 6.3 *Small group discussion develops listening and questioning skills.*

Conclusions

I loved doing this project because it was science based around talking, listening and questioning. I think my strategy of working in pairs and small groups worked well because it gave each child a chance to play around with ideas, test them out on each other and share them in a secure way. I think this made them feel responsible and in control, not just doing what the teacher had asked or set up for them to do.

As Barbara Lalljee (in press) says in *Talk across the curriculum*, 'working in pairs ensures that each child gets a hearing'. I also think that by starting from what children want to know about a topic, you can ensure that:

- the learning is relevant;
- it builds on what they already know and so the learning is more secure;
- it develops their language skills, which are basic to all learning and to human communication;
- it develops a child's sense of enquiry;
- science is seen as relevant, enjoyable and an essential part of daily life.

Working this way, I also found that I had actually covered much more of the curriculum requirements than I had set out to do in my term's plan!

Rosemary has shown how she uses her skills of language teaching to help her to improve her science teaching. When children are 'talking science' it does not only mean talking about science, it means they are using language as their way of doing their science. Language is the tool we use for observing, hypothesising, describing, comparing, classifying, analysing, etc. Language is a set of resources for making meaning and by finding out what the children were thinking and what they wanted to learn, Rosemary was able to use language to build on what they knew and to extend their thinking.

This chapter concludes section B, in which we have focused on a variety of teaching strategies in science. In each of these chapters, it has become clear that the role of language in the teaching of science is crucial, as is the need to find out what the child already knows and believes. Now we move on to section C and a collection of stories written by teachers about some of their science teaching. You will see in these stories some of the teaching strategies discussed in section B and how these are applied to children throughout the primary age range.

Section C Getting started

Getting organised

— 7 — Developing children's understanding of their bodies
Careen Leslie

In this chapter, Careen describes a unit of work in which she expands her pupil's knowledge about their body. She explains how drawing around the children's bodies on a large sheet of paper provided the opportunity for her to find out what each child understood about the inside of their own body. What was most fascinating was the children's notion of food being swallowed and somehow just floating around in their body. This work tells us a lot about how difficult it must be for children to understand what is inside their bodies and how they work. It shows how children's knowledge and understanding can be extended through a set of simple but effective activities.

Age	3–5
Situation	teachers with whole groups, small groups and individuals
Science	emphasis on children's intuitive understandings; questioning; exploration and investigation to introduce a scientist's view
Theme	children understanding what is inside their bodies and making sense of the processes of life.

Introduction

As adults, we have a reasonable understanding of our bodies and how they function. But what do young children understand about themselves? Do they make any sense of how they feel and how their body works? If they feel hungry, they know to eat something, but do they know where the food goes? Does it go down into their legs? This is a story of how I came to understand what my children knew about their bodies and how I moved their thinking forward.

I had twenty-nine three- to five-year-olds in my class. The teaching occurred over four weeks, on the Monday and Tuesday of each week, from 9.00 am to approximately 10.30 am. Each day, a whole-group time was followed by free-choice time. I used the following steps to explore this topic.

1 *Before views:* find out what children know.
2 *Questions:* find out what children would like to know.
3 *Explorations/Investigations:* teacher and children explore and investigate questions and jointly construct scientific understandings.
4 *After views:* record children's findings and compare with previous views.

What children understood about their bodies before

At a whole-group time, a bear puppet introduced the topic of 'Our Bodies' by asking children to tell the bear what they knew about their bodies. I drew all the body parts onto a white board, making up a funny-looking human. Interestingly, I saw later that many of the children had replicated in their drawings the spiky hair-do that I had created when I drew on the white board – the power of modelling!

The bear then asked the children if they knew what was inside their bodies. This was followed by writing up on some big sheets of paper everything the whole group knew about their bodies. The ideas the children had are shown in the concept map in Figure 7.1.

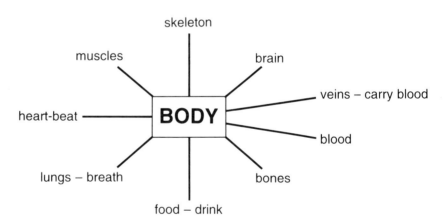

Figure 7.1 The children's concept of their bodies before investigations.

This process encouraged children to share their understandings and promote discussion about their bodies, both inside and out.

Free-choice time

I had some other adults in the class with me, and they helped trace around each child's body. Individually, the children then drew what they thought was inside themselves. This produced some amazing results (see Figure 7.2 overleaf). Their large body pictures have been redrawn on a computer (traced from a photograph).

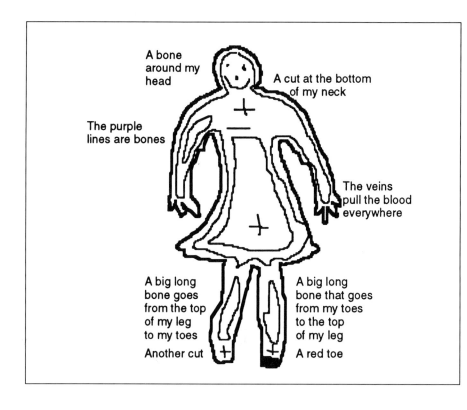

Figure 7.2 Catherine's concept of what is inside her body.

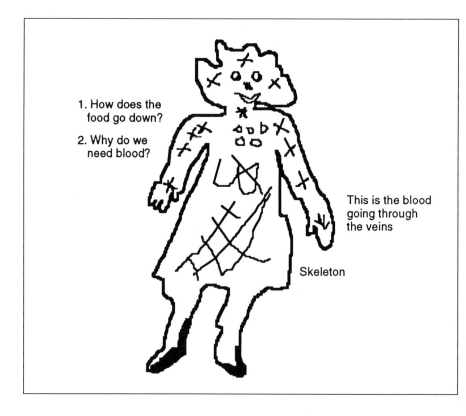

Figure 7.3 Tegan's understanding of what is inside her body.

Children's alternative views

Each day, I encouraged the children at whole-group time to share with each other what they had been doing or thinking. Children enjoyed sharing their understandings about themselves with the others. An example of this was when I asked four-year-old Jenny to explain what she had drawn and collaged inside her body. Jenny said that food travels down to your legs and that she had 'soft blood' in her toes. And when I asked the children if they knew why we have a belly button, there were some equally interesting answers:

> 'To help you breathe'
> 'Tummy button is in his brain.'
> 'I know where your heart is –
> behind your tummy.'

When asked what the brain did, one child provided the explanation, 'The brain wiggles yourself.'

Children's understandings of themselves

The following are some of the understandings that children freely shared at either whole-group time, or whilst they were drawing.

Cody: Blood is everywhere.

Joshua: When you bump your nose, blood goes everywhere.

Tegan: Veins carry blood around the body, and we have two holes – one for food and one for air.

Brenna: The skeleton is made up of bones.

Tegan in her 'before' views explained how blood goes through the veins. She even used different coloured pens to indicate the types of blood – a red and a blue colour showing the veins in her wrist. Tegan then went on to demonstrate her perception of the body's skeleton – this was indicated by crossed lines (see Figure 7.3). Tegan also thought we have two holes – one for food (even the food was drawn, as small round circles) and one for air. She had placed these two holes approximately where the lungs are normally. Another child, Cody, simply drew blood everywhere, using a red felt pen and many stroked lines (see Figure 7.4).

These views and understandings were not incorrect, but only partially formed, giving me a clear indication of what experiences to introduce to the children in order to help them develop a scientist's viewpoint and to begin to grapple with 'new knowledge'.

Questions

Once children had expressed their understandings about the topic under investigation, it was important to try and encourage them to ask scientific questions. I always find this difficult to do, and certainly research by Fleer (1991) has shown this to be problematic with very young children. However, I have found that, by encouraging children to ask scientific questions, I can find out what my group of children are most interested in investigating.

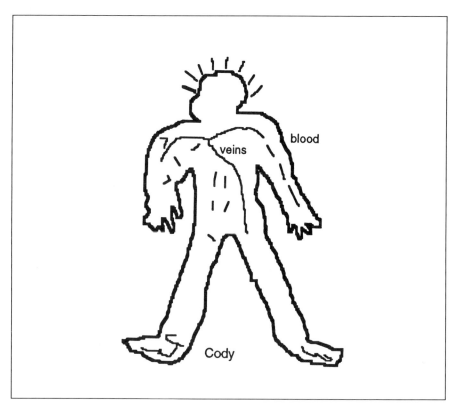

Figure 7.4 Cody's understanding of what is inside his body.

I explicitly introduced the idea of questioning by:

- modelling questions (by this I mean rephrasing their statements into questions and asking my own questions);
- having children ask each other questions at news/sharing time;
- having a child who had already asked questions to model them to the other children.

Tegan, aged five, naturally asked two questions when I traced around her body: 'How does the food go down?' and 'Why do we need blood?'. The next day, she was able to share these questions with the other children. This encouraged other children to ask two more questions: 'How does the skeleton go together?' and 'How do we breathe?'.

This session was followed by asking children where they would find out the answers to the questions. Some of the children's ideas were:

- go to a hospital;
- be dead – Tegan explained that 'another man could see what was inside the body' (this idea, which actually describes an autopsy, shows remarkable insight: as the teacher, I had to create the moment for her voice and opinion to be heard, as there were other children who said,

'That's not nice!', a common social value relating to death, and Tegan, being a shy girl, would not have continued to verbalise this idea);

- books;
- ask someone who knows more than you do.

Explorations/Investigations

At this stage, I worked together with the children to answer the scientific questions they had posed. I introduced models of the whole body system, including:

- a magnetic, layered body which enabled children to peel off the various layers of the body's system. A magic moment was when I slowly peeled off the layer of skin with one child peeking through her fingers in anticipation;
- a skeleton – a life-sized one and a small one for individual use at free-choice time;
- a life-sized organ model to give children a three-dimensional, representational figure to explore and investigate;
- real organs purchased from the butcher, such as kidneys and hearts.

Children excitedly moved from one table to the next, comparing and matching the various organ shapes using the different models, for example, the real kidney with the three-dimensional model.

The idea of the body being like a puzzle was introduced through discussion and hands-on investigations by the children. All adults worked closely with the children, scaffolding their understanding towards the scientists' view. Parents were also important participants in this learning process and welcomed the opportunity to 'visit' the insides of their bodies with their children. It represented a wonderful and powerful way for parents, children and teachers to become partners in the educational process.

The anatomy apron

To answer the specific questions 'How does the food go down?' and 'How do we breathe?', an Anatomy Apron (Educational Insights, 1986) was introduced. The Anatomy Apron is child sized and is worn by placing the apron over the child's head and tying at the sides. The apron clearly shows the children where the organs are positioned, both back and front. The organs (heart, lungs, liver, stomach, small intestine, large intestine and kidneys) are named and are easily attached to the washable, vinyl apron with Velcro.

Introducing the organ

At whole-group time, I asked a child to come out to the front and wear the apron whilst I introduced each organ. This gave children an overview of the body and promoted much discussion at free-choice time. Children thoroughly enjoyed matching the organ shapes and remembering each organ's name. Each organ was then introduced by its name, position in the body and its main function. Some examples of this were:

The heart Children were fascinated with the fact that the heart is like a pump and circulates the blood around the body. A group activity, which was fun for the children, involved children running very fast on the spot and then, in pairs, listening to each other's heart beating, using toilet rolls as stethoscopes.

A song we learnt
The heart is like a pump,
The heart is like a pump!
Pumping blood to help you grow,
The heart is like a pump!
(to the tune of 'Farmer in the Dell')

The lungs Children loved blowing up balloons and positioning them over the spot where their real lungs were, listening to the air going in and out. I asked questions such as:

- How are your lungs like these balloons?
- What makes the balloons get bigger?
- What makes the balloons get smaller?

I explained that when air is let out of the balloon it is called exhalation and blowing up the balloons was inhalation.

A song we learnt
Breathe, breathe, breathe in air,
And your lungs will grow!
In and out,
In and out,
That's the way it goes!
(to the tune of 'Row, row, row your boat')

The stomach This simulation exercise involved using a plastic bag, a banana, chopped biscuits and liquid as a graphic visual aid to help the children understand the process of the stomach muscles squeezing the food into a liquid, ready to move on to the small intestine. The children enjoyed watching one child modelling the apron, by pretending to eat the food, chew the food and squelch the food in the plastic bag until it no longer resembled a banana, biscuits and water. Children also heard too the sounds their stomach made by listening through a toilet roll.

The small intestine and large intestine For this organ, I demonstrated its function by pushing cooked oatmeal through an old stocking and showing how it winds up tightly inside the body. The analogy of 'it's as long as a tall tree' was well remembered by the children and enabled them to grasp the wonder and complexity of the body. The understanding of how waste is eliminated from the body's system could now be easily discussed.

Other activities used to make connections throughout the classroom were:

- a felt story using the internal organs;
- songs for each organ;
- puzzles of the body;
- a collage of matchsticks to make a skeleton;
- drawings of children's ideas.

What children found out

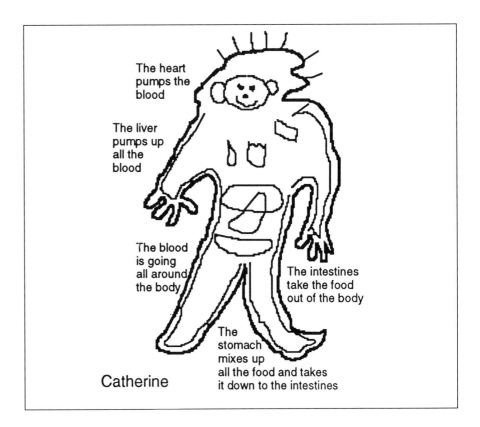

Figure 7.5 Catherine's drawing of what is inside her body.

As can be seen by Catherine's drawing (Figure 7.5), she had grasped the shape, position and function of the organs inside her body – understandings that were previously not known.

An 'after view' group concept was drawn (Figure 7.6) which highlights the children's understanding of the body as a system rather than as a collection of one-off organs.

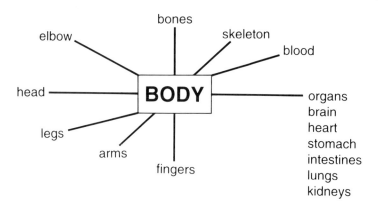

Figure 7.6 *The children's concept of their bodies after investigations.*

The children's bodies were once again drawn around and compared with their 'before' views. For example, three-year-old Ruben, who started his first day at the centre when this 'Body' unit began, could at first only manage to paste a few collage items on his trace-around body. Compare this to the 'after' view, where he could not only draw some of the organs, but also name and explain some of them (see Figure 7.7).

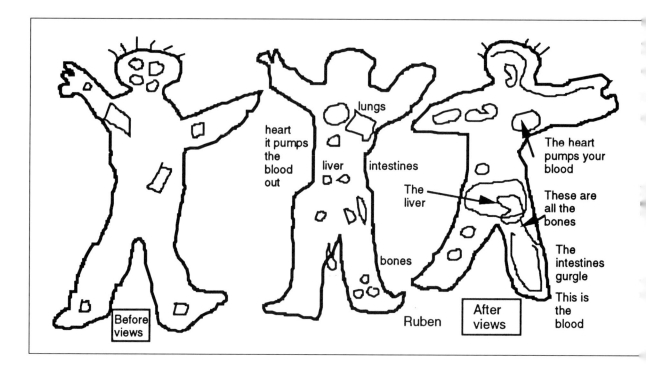

Figure 7.7 *Three line drawings representing Ruben's three body outlines. These outlines show progressively how his understanding of what he thought was inside his body developed over time.*

Conclusion

This whole teaching sequence, I believe, enhanced the children's understanding of what is inside their bodies and empowered them to discuss everyday events using scientific language and at a greater depth than before. For example, at fruit time, the children discussed what was happening to the fruit they were eating. When they fell over and cut themselves, they knew it was the liver that was helping from the inside as well as the Band-Aid on the outside. When they ran, their heart was pumping blood very fast throughout their body. When they ate too quickly, they often would have a stomach ache.

By following this modified interactive approach to teaching science, children were challenged beyond normal expectations and encouraged to experience the joys associated with scientific discovery.

What this chapter has shown is the importance of eliciting children's thinking before we start teaching. Without the activities undertaken here, any teaching strategy would be based on a guess about children's understanding and likely to be inappropriate. The respect for the children's ideas running through this piece helps to draw children into the process of learning. So while as teachers we always have to hang on to a clear idea of where we are going, these methods help to tell us where we should be starting from.

— 8 — Putting the green back into plants!
Vicky Bresnan

We can tend to get a bit repetitive when we teach children about plants, but we included this chapter because it represented a different approach. Here, Vicky describes a unit on plants for four- and five-year-olds that is broad enough to run for many weeks, and yet sufficiently focused for the content to be based upon the questions asked by the children. Through actively trying to find out what her children wanted to know about plants, five areas were identified for exploration: seeds, flowers, fruit, trees and what plants are used for. The framework for the unit, planned around her children's interests, was probably much more comprehensive than one she would have imposed upon the children. Whilst there was a lot of teacher input, it occurred only after Vicky had found out what the children already knew about plants and what they were interested in finding out.

This chapter also touches on the interesting issue of deciding whether an object is alive or dead. Pursuing this question allows the processes of life to be explored. The work also shows how a concept map can be used as a starting point for work based on children's existing knowledge.

Age	4–5
Situation	teacher working with whole and small groups
Science	plants
Theme	finding out what children know and helping them to answer their scientific questions

Introduction

Joel wanted me to put the green back into the plant that had been deprived of light. I asked him how he thought we might do that. Monique said, 'put it in the sun, and the green comes back'. This conversation is an example of what happens when I organise science experiences for my children.

The approach I take in teaching science is one in which the children are able to have some input into the content of the theme. This is done in a number of ways. Firstly, we construct a concept map, which enables me to collect information about what knowledge and concepts the children already have acquired about a given topic (Figure 8.1). Then, after some investigations which stimulate further curiosity, the children are given opportunities to ask their own questions and investigate the answers.

A concept map is put together by listing all the knowledge, beliefs and attitudes that the children have about the area being studied. This is a sharing time for the whole group and can be used to motivate discussion about a topic. By using this approach the children:

- learn from each other;
- decide their own areas of study;
- extend existing understandings.

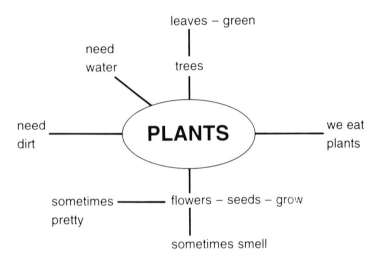

Figure 8.1 A group concept map of plants

This approach also gave me the starting point for planning further experiences for the children. In addition, I was able to learn a lot more about the children themselves, what they know, how they learn and what types of learning experiences they have already participated in. Although knowledge is important, I see the skill of how to attain knowledge as being equally important. Once children have the skills of learning, knowledge will never be unattainable. My own experience as a teacher has shown me that children learn and retain a lot more from each experience if they have been an active participant in all areas, from pre-planning to evaluation or reflection of a topic.

My role was to assist the children in setting up learning experiences which would move their thinking forward. We used videos, books, audio tapes, film strips and experiments to encourage the children to absorb information, sort and classify, observe, and then to make generalisations about their research. The children also participated in excursions, which helped them to make use of other resources.

The topic of 'Plants'

The topic of 'Plants' is one that we have spent time working with on several occasions during the year. The children participated in setting up a vegetable patch from scratch and harvested their first crop earlier this year. We set up flower beds in old tyres as part of a conservation programme, and the children participated in beautifying an eroded hill near the sandpit. In addition, the children have spoken about plants in the context of seasonal changes, observations of which were quite natural in our school environment, as the garden contains both deciduous and evergreen trees.

Starting another project on plants gave everyone an opportunity to establish what we had already learnt and to extend existing knowledge. There was no need for a motivator, as the children check their vegetable patch daily.

From the concept map, I was able to determine five main areas of interest:

- seeds: where do they come from?
- flowers: why are there flowers?
- fruit: why do plants have fruit?
- trees: what are trees?
- what are plants used for?

The overall focus questions were:

1 How do plants grow?

2 Why do we need plants?

The idea was to help the children to find information, make observations and conduct experiments to answer the question 'how do plants grow?'. When we started the topic, I also told the children that we would talk about the life cycle of a plant, and some of the children asked 'what's that?'. We were then able to transfer our previously acquired knowledge about the life cycles of ants, caterpillars, spiders and frogs to our new topic of plants. The classroom dialogue was as follows:

Teacher: What do you think life cycle means? . . . Well, think back to when we talked about butterflies. What was the life cycle of a butterfly?

Sally: The butterfly lays eggs and the eggs turn into caterpillars.

Teacher: Then what happened?

Jessica: The caterpillar eats and eats . . .

Teacher: The caterpillar eats and gets fat and makes a . . . ?

Jessica: A chrysalis.

Teacher:	What happens in the chrysalis, Jessica?
Jessica:	A butterfly.
Teacher:	Then what happens?
Jessica:	Butterflies mate to make eggs, and the eggs grow.
Teacher:	What happens to the butterfly?
Jessica:	They fall down dead.
Teacher:	We've learnt about other life cycles, too, which tell stories about how animals are born and live and die. What do you think life cycle means?
Jessica:	Plants living and dying.
Teacher:	Yes, plants living and dying, and also making new plants, and that's what we want to find out about. How do we do that . . . find out about plants?
Jessica:	Ask people.
Teacher:	Who might that be?
Jessica:	People who know about plants.
Teacher:	Good suggestion. What else can we do?
Sally:	Books.
Teacher:	What about books? Tell me with more words about books.
Jessica:	We can look into books.
Teacher:	Good idea. What else can we do? . . . Well, perhaps we can think of some other things we can do as we go.

This dialogue set the scene and also showed me that the children hadn't yet internalised the other types of experiences we had had when trying to learn about other topics.

At this point it was necessary to create a learning environment where learning will be natural and safe when children are immersed in language. In this case, it was the language of science and plants. Figure 8.2 shows the range of experiences across the curriculum used to explore the area.

The group began investigations after formulating a focus question, for example, 'How do plants live?'. Not all children participated in all investigations. However, the discussion, reflection and recording of our findings were done in group situations.

Seeds

Seeds was the first topic we talked about. We read a book on seeds, and the children were then given an opportunity to investigate books at the reading table.

The children sorted seeds and categorised them according to size and then colour. We made a seed chart as a group activity. Some children were surprised that one tiny seed could result in a huge tree.

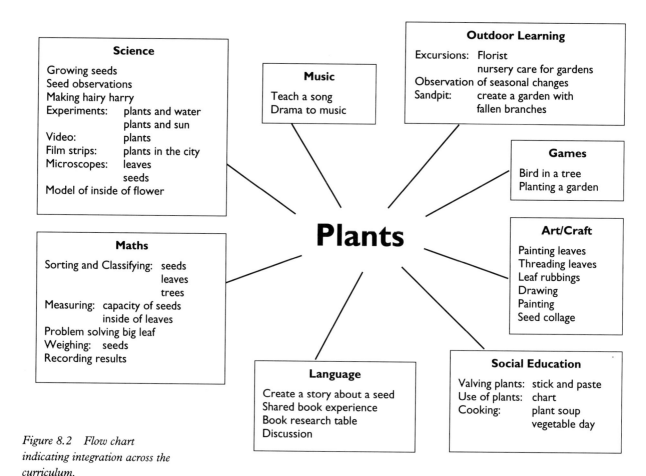

Figure 8.2 Flow chart indicating integration across the curriculum.

This also led to a discussion on how many seeds there are in an apple, an orange and a mandarin. Kiera wanted to know which fruit has the most seeds, so we counted seeds. What was interesting, though, was that the number of seeds varied each day. Kiwi fruit and tomatoes had more seeds than we could count.

We looked at the insides of seeds through a microscope. We also looked at the outside of a seed.

We planted broad beans, and the children watched as a root system developed and grew downward, while the stems and leaves grew upward. Both came from the same hole/black spot on the seed. William made these observations and always put in a soil line on his drawings of plants.

We made a display of the ways in which seeds disperse. We collected the information from books, film strips and our excursions. The children learnt that seeds disperse:

- as a pod explodes?
- by the wind blowing them?
- by birds eating and excreting them?
- by animals carrying them in their fur?

- by fire breaking the pod open?
- when we plant them.

On a film strip, we saw plants that were growing in peculiar places, such as roof gutterings and cracks in concrete. This led to a discussion on how the seeds got there. Some children were able to determine this logically, while others created some fascinating stories.

William wanted to know how plants 'got planted before there were men to do it, and where did the first seed come from?'. It was easy to talk through the first part of the question by discussing the findings of our seed dispersion display. By a process of elimination, William was able to work out for himself how seeds were dispersed. William decided that seeds could be carried by dinosaurs, pterodactyls, wind and water. Some may have exploded as well.

Flowers

After seeds, the children talked about the various parts of plants – flowering plants – and we looked at the second part of William's question, where seeds come from.

Some of the children drew their findings whilst others stuck the parts of the flower onto paper. We showed the tubes (the zylem) inside the plant by dying the water and allowing the flower to absorb the water, with the result that the flower changed colour.

Kiera alerted the group to the veins in the petal after observing the petals had thin lines of colour. We looked in books to determine which parts of the flowering plant had veins. Our conclusion was reaffirmed in a video entitled 'How plants grow'.

When pulling the petals off a carnation one by one, Joel was excited by the discovery of the seed. We recorded our findings in a narrative story that the group wrote together.

OUR FLOWER STORY
Once upon a time, there was a red flower. It had a long stem and some green leaves. It had roots under the ground to hold the plant straight. The roots sucked up the water and took it to all the leaves and petals in the flower. The flower made its food in the leaves. The food was made of sun and water and stuff from the dirt. The flower was happy when the bees visited. They put yellow dust called pollen into the flower. The petals fell off the flower and there was a seed.

The children played with the story on the felt board which showed me that they understood the flower story. Some children also drew their understandings and labelled the parts of the flower. As in many aspects of early years education, 'story' was central to this unit of teaching. It was as the children explored their 'new knowledge' through the language of the story that I could see what they understood.

Fruit Cooking days enabled us to talk about the different parts of plants. We categorised vegetables into groups, according to the parts we eat.

Leaves	Bulbs and roots	Fruit	Flowers	Stalks
spinach	potato	capsicum	cauliflower	celery
lettuce	carrot	tomato	broccoli	
	turnip			
	swede			
	onion			

Living or dead?

Throughout our research, there was an underlying question which remained unanswered but which we talked about at great length: 'What is living and what is dead?'

We planted some daffodil bulbs, which were slow in coming up, and this generated some concern and questions from the children as to why the bulbs didn't grow. After discussion, the children came up with the following answers:

- 'They were bad bulbs.'
- 'They were dead.'
- 'There was something wrong with the dirt.'

Ben, aged four, suggested that we 'dig up the bulbs to see if they are still in the dirt'. I was able to avoid such a drastic measure by suggesting that we give the bulbs some extra time. Then, if the bulbs hadn't made any progress, we could dig one up to see what may be wrong. The children decided on ten school days, and we started a calendar to mark off the days. However, we forgot about the bulbs, because we were distracted doing other things. Then, four weeks later, a parent involved in the planting of the daffodils noticed green shoots coming through the ground and alerted her child. Kelly got everyone excited about her mother's discovery. So, what happened to our bulbs? Nothing – we just had to give them more time.

'What does dead mean?' 'When is a seed dead?' These questions arose because the children felt that was the explanation for the delay in our daffodil garden. Because the bulbs *did* grow, the children then declared that they were indeed alive. However, I showed them a bulb from another plant and asked them what they thought.

Jackie: It looks dead. The skin is wrinkly and crackly.

Teacher: Does that mean that it's living or non-living? ... How could we find out?

Jackie: Put it in some dirt.

Teacher: What does that do?

Jackie: Makes it grow.

I then showed them all sorts of seeds, leaves, fruit and vegetables and asked if they could tell me if the plant was living or dead. We made a list of what we thought.

Living	**Non-living**
carrot	autumn leaf – dried and brown
potato	seed – like a rock
apple	stem – no leaves
flower	twig – broken and brown

The children based their classification on the colour and appearance of each item. It was clear to me that the children did not have a clear understanding of what was living, so I needed to set a task to illustrate the difference. One of the messages we received whilst on the excursion was that flowers are no longer living once they are cut or picked. We began our study with a plant, grass seeds and a question.

Teacher: How do we know this is living?

Jacob: It's green.

Jessica: Smelly.

Rowan: Juice is in the middle.

I then asked the children to tell me about a cut flower. Is this flower living or non-living? This question created a discussion, so we recorded the results.

Living	**Non-living**
It's yellow.	It's not in dirt.
The stem is hard.	There are no roots.
The flower is pretty.	

Since there were two differing opinions, we agreed to talk about the flower the next day. I purposely left the flower on the table and asked the group the following day to describe what happened to the flower; I received the following answers:

- 'It won't stand up.'
- 'The petals are soggy.'
- 'The flower is dead.'

The children were then able to classify the cut flower as non-living.

I presented an orange to the children and asked them to think about the flower. I asked them to place the orange into the group they thought was appropriate. Most children chose non-living because the fruit was not part of the plant.

The evaluative activity was to construct a chart of living and non-living objects in picture form, as a record of our discoveries.

The most exciting aspect of teaching is the amount of follow-up that is pursued at home. Children are constantly bringing their discoveries to school. For example, Stephen brought a mushroom that had been left on a paper towel overnight and had made an imprint on the paper. This is where the mushroom had dropped the spores that it relies on for reproduction. This opened up another investigation: the life cycle of a mushroom.

Whilst talking about what makes plants grow, one child said, 'plants need water and dirt to keep healthy'. I asked the group to predict what would happen if we took water, light and both water and light away from a plant. Kiera thought out the experiment; using four cups filled with dirt, cress seeds for each and symbols for each cup (see Figure 8.3).

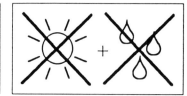

Figure 8.3 Symbols used for labelling experiments

The results were not as I would have expected. All the plants grew, but in different ways. The results were recorded, and the children decided that, for a plant to be green, it needed light, and for lots of stems, it also needed water.

This unit of work was now winding down, but we would revisit it later that year when we looked at spring and how spring affects plants. We also had a guest to teach us how to make paper, a by-product of plants. Later in the year, we would be going to the Botanical Gardens to make further observations involving all our senses . . . but that's another story.

This work shows how a group concept map formed the basis of the teacher's planning, and the ways in which the teacher used the children's questions as a basis for teaching. It also shows that flowers are not dead – the study of them, that is! The impression is one of interested children generating good questions and constructing sensible investigations to find answers – exactly what science is doing in the adult world.

9 — A moving experience
Marita Corra

The topic of movement is not only something very familiar to children, since it is something they do all of the time, but one that lends itself to some very interesting scientific investigations. In this chapter Marita describes what she did with her children on this topic and the children's responses are fascinating. Marita has captured the connections that children made as they think through the concept of movement as well as how she went about moving their thinking forward.

Age	4–5
Situation	teacher with whole and small groups
Science	exploring children's understanding of movement
Themes	what children understand about 'movement'; expanding these ideas; questions and investigations

What is movement?

Movement is basically changing from one place or position to another. Many things move and there are many forms of movement in both animate and inanimate objects.

- change of position or posture;
- going from one place to another;
- motion or a series of combined actions;
- changing one's place of residence;
- moving parts of a mechanism.

The list could continue indefinitely to include music, poetry and even political groups! Movement is prominent in our daily lives – most things around us tend to move in some way, sometimes visibly, sometimes not.

The topic of 'movement' provides experiences in scientific observation, investigation, classification, problem solving and recording. Children start with preliminary discussions that explore their understanding and knowledge of the topic. Relating movement to their daily life, they go on to pose questions and plan investigations to explore their views, challenge ideas, present alternatives and draw meaningful conclusions. Finally, they discuss and record their findings, thinking about whether their ideas and views have changed and what knowledge they now have on the topic.

Where to start?

To get the topic started, I used the following:

- a short aerobics session with the children each morning;
- books related to movement;
- a large display of 'doing' words;
- a discussion in small groups;
- a concept map on 'movement';
- the focus question: 'What do we know about movement?'.

Figure 9.1 Children's concepts of movement before investigation.

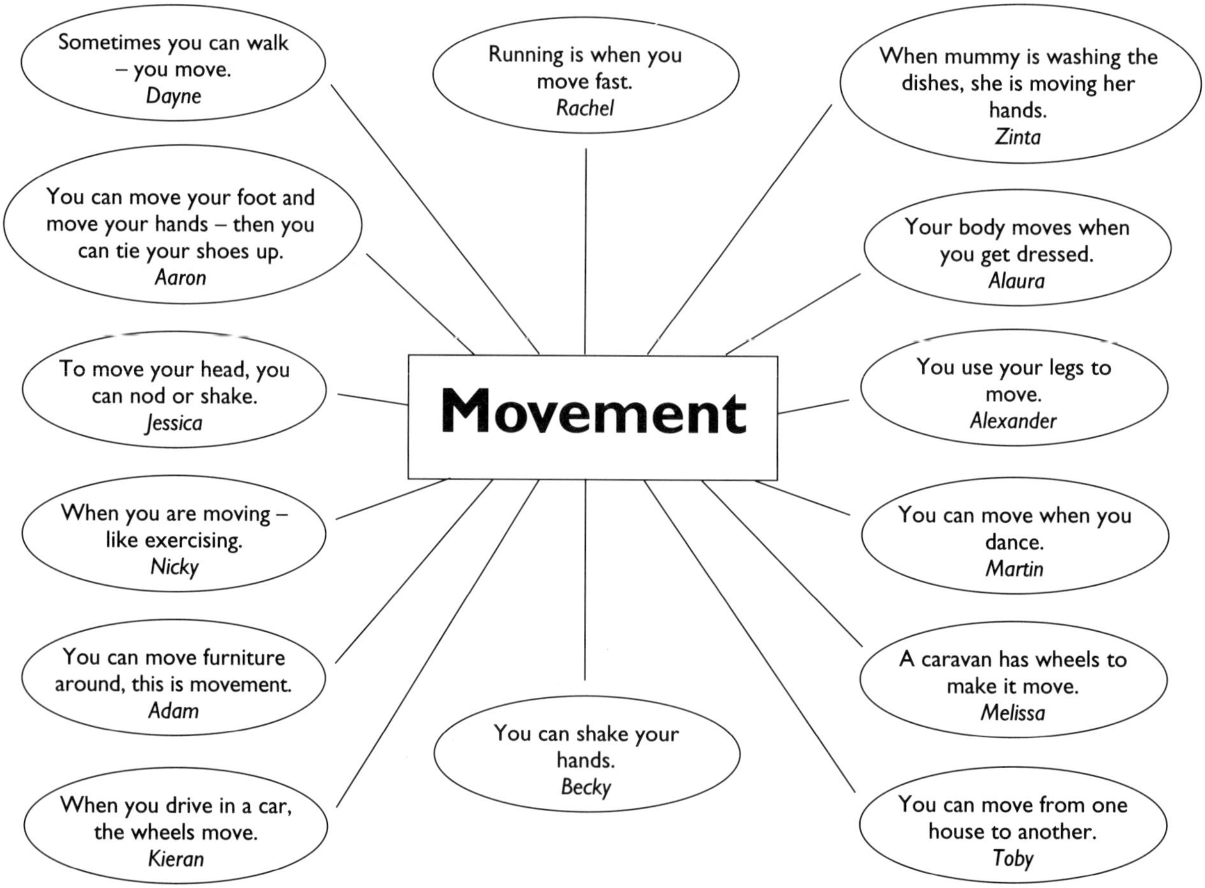

Movement

- Sometimes you can walk – you move. *Dayne*
- Running is when you move fast. *Rachel*
- When mummy is washing the dishes, she is moving her hands. *Zinta*
- You can move your foot and move your hands – then you can tie your shoes up. *Aaron*
- Your body moves when you get dressed. *Alaura*
- To move your head, you can nod or shake. *Jessica*
- You use your legs to move. *Alexander*
- When you are moving – like exercising. *Nicky*
- You can move when you dance. *Martin*
- You can move furniture around, this is movement. *Adam*
- A caravan has wheels to make it move. *Melissa*
- You can shake your hands. *Becky*
- When you drive in a car, the wheels move. *Kieran*
- You can move from one house to another. *Toby*

Following the discussions, I used a concept map to list the children's views and ideas. These are shown in Figure 9.1.

Children's views on 'movement' were quite predictable. The majority of them associated it with body movement and position.

Becky: You use your legs to move.

Alexander: To move your head, you need to nod or shake.

Other children associated movement with travel. They discussed the importance of wheels in getting a car to go from one place to another. I was quite surprised when a child linked changing residence with movement. This concept of 'moving' was, according to the child, the 'same thing as movement'. Following this conversation, another child decided that moving was associated with movement.

> You can move furniture around a house and change things in a room. This is called moving.   Ceili

All the children were confident about sharing their views within a large group. Some children were prepared to give examples and even justify their comments. This showed a real commitment to their ideas and views of the topic.

Explorations

The children spent time exploring their ideas. We played games such as 'Punchinello', 'Follow the Leader' and 'Simon Says'. The children used their bodies in a variety of positions and manner. Such games were non-threatening, lots of fun and provided a means of learning more about how the body moves.

The children also made a graph of how they came to pre-school: walking, riding bikes or in a car. We then transferred this information onto paper so the children could see the numerical representation as well as the forms of movement (Figure 9.2).

Figure 9.2 How did we come to pre-school?

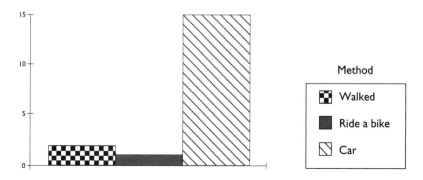

Children's questions

The children in the group were familiar and at ease with questioning. After the exploratory activities, they divided into small groups of between six and eight to pose questions for further investigations and exploration. The children were asked:

- What would you like to find out about movement?
- What investigations could you do to help you understand more about movement?

The children's questions were very focused and they managed to come up with some unusual ones! Their final list of questions was as follows:

1. How do wheels move?
2. How do our feet move? (This question was extended to include other parts of the body.)
3. Why do we get dizzy?
4. How can we put clothes on?
5. How does a windmill move?
6. How does water move along pipes?

Investigations

Question 1: How do wheels move?

Children believed that the purpose of a wheel was to 'turn round and round'. In order to move, a wheel had to turn or spin. The children investigated this question with a variety of materials and experiences. Initial investigations with a train set and constructions sets enabled the children to observe the role and function of the wheel. Outdoors, bikes and scooters were also used.

Rachel: If the wheels were square, they wouldn't turn.

Nicky: Wheels move because they are circling all the time.

Zinta: A factory only makes wheels in a circle shape anyway!

Further investigations inside included a variety of wheels on a table, where they investigated the problem of how wheels can move by themselves. With no adult support, the children collected boards and made themselves a slope.

Aaron: A wheel will move on a slope, it rolls.

Melissa: A wheel, that is by itself, not on a car, will only move if the wood is tipped up a bit.

The children then went to observe cars and trucks passing the school. This lead to a closer examination of a stationary car. What did they discover?

Adam: The wheels are joined together with a big pipe. I bet it somehow has something to do with the wheels, making them all turn together.

A factual text and car manual gave the children information about axles. Budding mechanics then made their own vehicles with boxes, wheels and, of course, axles!

Question 2: How do our feet move?

The children's initial views linked body movement to the brain and bones (see Figure 9.3).

Rachel: It's the brain that makes you think, and it tells your feet to move.

Jacob: If you had no bones, then you couldn't move your feet.

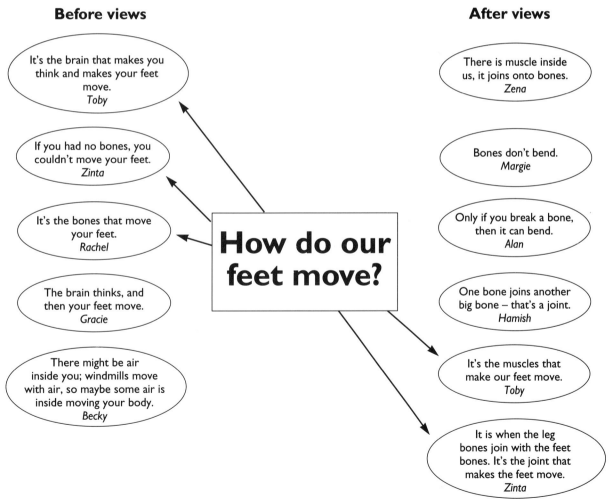

Figure 9.3   The children's ideas of how feet move before investigation.

The children showed a definite understanding of human physiology and were comfortable discussing the functions of body parts and organs. I therefore rephrased the question to accept their views and to develop further ideas, for example: 'What is inside your body, apart from bones and a brain, that makes your feet move?'.

> There is air inside you; windmills move with air, so maybe it is the air moving around inside you that can make you move. Dayne

Who can argue with such logic? One child volunteered a parent as an 'expert' to answer the question: 'My mum will tell us because that's her work'. The expert was a physiotherapist who was able to bring models of legs, shoulders and spine plus resource and picture books. In a large group talk, she introduced muscles and joints. She shared her knowledge and encouraged children to extend their understandings of their body. Reflecting on the visit by the physiotherapist, the children had acquired new information.

Rachel: Only if you break a bone, then it can bend, bones don't bend.

Aaron: There is muscle inside us, it joins onto bones. It's muscle that makes our feet bend up and down.

Martin: One bone joins another bone, it's a joint, it's an ankle.

Question 3: Why do we get dizzy?

This question is linked to the previous question, with its answer lying in human physiology. The children's initial answers were impressive!

Zinta: It's because when we turn round and round and round, our bones get all twisted round. That's why we are wobbly when we stand up.

Becky: It's really the house that is turning around us, it's going round and round, and *it* is making us dizzy.

The children connected the dizzy feeling with the motion of spinning around. When asked who the most appropriate person would be to help us with out investigations, the common response was 'a doctor'. Indeed, the answer proved to be quite complex, involving medical terms and description; the doctor was able to describe the function of the ear to the children and shared her knowledge, with the help of visual aids, of how turning round makes the fluid in the ear spin too. The children absorbed all her explanations and were quite fascinated, but I still think it was much easier for them to believe their twisted bones rather than their ears make them dizzy!

Question 4: How can we put clothes on?

Following the visit of the physiotherapist, the children were able to answer this question. A large group discussion focused on the types of movements we make when getting dressed. The children acted out incidents, for example:

Toby: You need to bend your knee and lift up your foot to put on socks.

Melissa: You need to roll your shoulders backwards to put on your coat.

Games and songs, such as 'Simon Says' and 'This is the way we dress ourselves' were played. In the Home Area, children spent time 'dressing up' and dressing dolls. This encouraged lots of talk and sharing of movement words.

Question 5: How does a windmill move?

All the children could clearly describe the function and role of a windmill (see Figure 9.4), even though not one child had ever observed or been in close proximity to a windmill! Asked about the origin of their knowledge, television and books were given as the sources of their information.

Cameron: The windmill is shaped like a large daisy flower, the shape is very important.

Ceili: The triangles catch the wind and turn round and round.

Rachel: If it wasn't a windy day, the windmill wouldn't move.

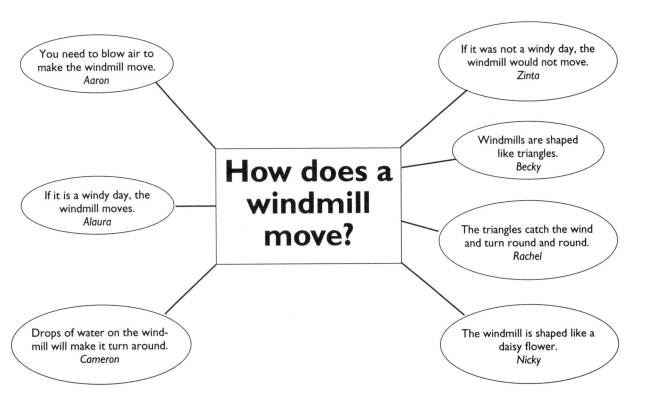

Figure 9.4 The children's concepts of the function and role of a windmill.

Our investigations focused on texts that described windmills, and this served to clarify ideas and give models for a craft activity to make windmills. One child was fascinated by windmills' sails and after making his own model, came up with a further question to investigate: 'Can drops of water falling on the windmill make it turn around?'.

Question 6: How does water move along pipes?

In a previous unit of work, the children had spent time observing and investigating pipes. A plumber had supplied information on the design and operation of pipes. In addition, a popular indoor and outdoor activity is pouring water through pipes. The children therefore had direct, concrete experiences from which to draw ideas and views. They set up investigations which included various lengths and sizes of pipes and beakers of water. I left the children to observe and discover the effects for themselves. It was interesting for me to observe the similarities between this investigation and the investigation with wheels. The children used the same principles and rules in each experiment.

Grace: If the pipes are sloped this way *(vertically)* and not this way *(horizontally)*, the water runs through really quickly.

Adam: The pipes need to be sloped for the water to move along.

Following on from this, I shared my knowledge of pumps and rural water supplies, where water can be stored in pipes and moved along with water and rain.

Reflection

This unit enabled all the children to understand the importance of movement in their daily lives. They were able to expand their knowledge of body movement and understanding of human physiology. The children also explored natural phenomena, looking at cause and effect of the movement of water and air. They also investigated the performance of wheels and tried to imagine what life would be like without wheels.

Rachel: No wheels – no cars, bikes, scooters, buses or even wheelbarrows – yuk!

Throughout the investigations, the children were continually applying the scientific skills of observing, inferring, classifying, predicting, measuring and communicating. By the end of the unit, all the children were able to formulate hypotheses on movement.

Becky: Movement is any type of moving!

This chapter shows the importance of talking as children learn science. Marita has listened intently to what her children have said and she works with them to build up their knowledge. It shows how, even with nursery children, it is possible to do work that will improve their scientific knowledge and understanding. Helping children to make sense of their everyday experiences is something we can do from the very beginning of children's school lives.

10 — Straight bits and bendy bits: an exploration of paper bridges
Lindsey Weimers

We have included this chapter because of the way that it describes the process of a few children learning by what they do. Lindsey also highlights the everyday problems of when, how and how much to intervene. These young children's use of books for information is interesting, and their eventual explanations of what happened and why are fascinating. Forces are often difficult to teach as you can't see them, but this is definitely a success story.

Age	6
Situation	teacher with whole class, groups and individuals
Science	the forces involved when building strong structures
Theme	exploring some of the factors involved in making strong bridges out of paper

Introduction

My class of six year olds were following a topic on 'Building'. I had planned for a variety of cross-curricula experiences, and one of the scientific areas I wanted the children to explore was that of 'Forces'. In my initial planning, I had anticipated this arising in a variety of possible ways, such as an exploration of levers and pulleys when looking at cranes and other equipment used on a building site, or when exploring which factors were involved in making buildings strong. This is an account of one particular investigation.

The class is organised into four groups of seven or eight children who alternate at various activities. One set of activities is called 'Investigating', and here the children are used to having a problem to explore as they choose, either on their own or working with other children.

At the start of every topic, the children are involved in brainstorming ideas for exploration. One of the questions that arose out of this was

'How do you make a strong bridge?'. I decided that a helpful way to explore some of the issues involved would be to give the children the challenge of trying to build a strong bridge out of paper.

The investigation: can you build a strong bridge out of paper?

We began by discussing this problem as a whole class, both to enable the children to share ideas and to help me assess where they were in their understanding of the factors involved.

The initial suggestion from one child was to build two Lego piers and to balance a piece of thick paper across them. Several hands went up to complain that this would not be very strong. I asked if anyone could think of a better way; I anticipated the complete rejection of this approach and some completely new suggestions, but I was wrong. None of the suggestions put forward improved on the basic idea of a flat piece of paper balanced on two piers. However, there were a variety of modifications. Matt thought that the Lego piers would need to be quite close together. Sara thought that there should be three piers. Graham thought that a block should be placed on each end of the paper to hold it in place on the Lego. Jivan suggested finding a long strip of Lego to place under the paper.

We also talked about how we would know it was strong. Most children thought we should put something on top of the bridge. Some children suggested testing the bridges by seeing how many bricks they would hold. So with these initial ideas, the children began the investigation. The Investigating Corner is permanently resourced with a range of construction material, and the children have access to drawers with a variety of other resources, such as sticky tape, string, metal fasteners, plasticine, etc. I provided a tray of paper in assorted sizes. The thickest was coloured painting paper.

First attempts

I was interested to see that all the children in the first group kept to the initial idea of two piers with paper stretched across it. Most children chose to work with a friend, and much of the time was spent making fat Lego piers to hold the paper. All the children found that, when they placed paper between the piers and tried to balance Lego on top, their bridge would collapse after two or three bricks at most had been added.

I asked them if they could think of a way to make the bridges stronger. Two children made a third pier and placed it in the middle of the paper bridge. They then managed to get several pieces of Lego to balance on the bridge, much to their delight, but only because the pieces were resting on the pier rather than the paper itself. I hated to dash their enthusiasm! I asked them what the bricks were resting on; they looked and noticed it was the Lego. They moved the bricks to the unsupported paper, and it collapsed immediately. They tried again, but did not change the basic design.

The second group also continued using the idea of a flat piece of paper balanced between two piers. All the bridges were very unsuccessful, except Harriet's. She used sticky tape to fix the two ends of the paper to the Lego, which helped to keep the paper in place when she tested it with bricks, but did not prevent the two piers eventually falling inwards with the weight of the load. By the time the two other groups had completed their first turns, no real design improvements had occurred, and the children were getting fed up with their unsuccessful attempts.

New solutions

I was now faced with the common teacher dilemma of how far to intervene in the process. I felt that many of the children were ready to grasp new ideas, but could not yet imagine other possible approaches by themselves; to use Vygotsky's terminology, they needed some 'scaffolding'. I therefore produced a range of pictures and books related to bridge building and introduced them to the whole class for discussion.

The initial idea that sparked new processes was that of a girder: one book had a picture of some bridges made from card folded in various ways. We did not at this point discuss the issue any further. I left the first group to try again on their own, leaving the books and pictures as reference material.

The children in the first group began by folding their pieces of paper just once, and immediately noticed that the structures were stronger and could hold more bricks. Some children were content with this achievement, but others wanted to make theirs even stronger. Edward tried two folds at the side and found he had increased his load carrying capacity from thirty-two bricks to forty (Figure 10.1).

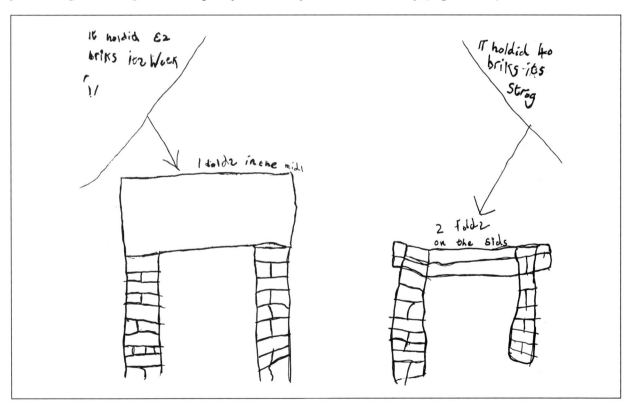

Figure 10.1   Edward's use of folds to strengthen his bridge.

Jenny's bridge

Figure 10.2 Jenny's first attempt at using folds.

Figure 10.3 Hannah's bridge.

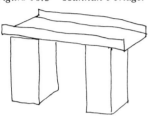

Figure 10.4 Jenny's second attempt.

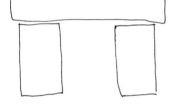

Jenny was not very confident about her work. She often looked worried when experimenting with new ideas. Although she was surrounded by other children making bridges, she was concentrating solely on her own rather than looking round as some of the other children were doing. She came to me very disappointed, as her bridge would only hold four bricks. I found that she was still making the bridge without any folds, just using a straight piece of paper between two piers. I asked her if she could think of a way to make it stronger. She looked worried and said nothing, but then looked round at the others. 'I could bend it,' she said, rather tentatively. She tried, but instead of folding the paper longways, she folded the two ends down over the piers (Figure 10.2).

When she tested it, she was surprised to find it could only hold three bricks. I suggested she had a look at some other bridges again. She watched Hannah testing her bridge, which had long upturned edges (Figure 10.3).

Jenny tried using this idea and was delighted to find it would hold twenty bricks. She then thought she would try another way (Figure 10.4).

However, this was the least successful and held only two bricks before the paper fell down. When I asked Jenny what she thought she had found out, she said, 'My other bridge only holded two 'cos I folded it here (*she pointed to the two short ends*) and it's still bendy but I folded this one here (*she pointed to the long edges*) and it's strong here and holds the Lego bricks up'.

Figure 10.5 Owen's drawing of his bridge with Lego piers.

Owen's discovery

The second group began like the first to make bridges strong by folding the paper. But Owen chose to look at the books before starting. He suddenly came running up to me very excited: 'Look, this ones got a bendy bit underneath'. He was pointing to a picture of an arched bridge, where you could see the arch clearly supporting the flat top section. I suggested he tried to make one like that. He used two bits of paper. He placed one piece bent between the two piers and placed the other flat on top (Figure 10.5). It held seven big bits of Lego. He was very pleased and showed his friends.

The arch idea spread quickly, and by the third group's turn, everyone was trying out arches in different ways. Anna tried making two complete tubes under the top surface. It held seventy-five bits of Lego (Figure 10.6).

Figure 10.6 Anna's use of arches enabled her bridge to support seventy-five bricks.

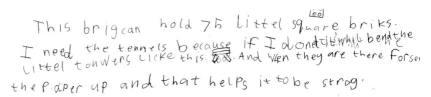

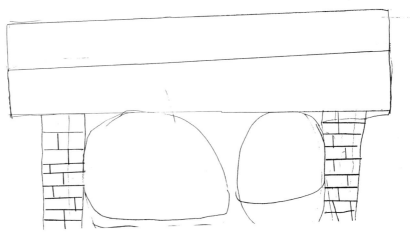

James made his arches with pointed tops and used three to support his top section of paper. This also proved very strong and held seventy-five bits of Lego. James remembered what we had been finding out in our shape work and correctly described his arches as 'like triangular prisms'.

Is it fair?

As the bridges became stronger, the children became very competitive about the number of Lego pieces each would hold. However, only a few children were concerned that some Lego pieces were bigger than others! If it held seventy-five, then that was all that bothered them, even if twenty larger bricks would have weighed more. I asked two children using different-sized bricks which they thought would weigh the most. They readily decided the bigger bricks were heavier, but although they then tried to use similar bricks, the complexities involved in keeping this variable constant soon defeated them and interest in the number involved again took over. By the end of the investigation, only five or six children were really concerned about the fairness of the testing and worked hard to keep the size of bricks constant when comparing bridge strengths.

What did they learn?

At the end of everyone's third session of bridge investigations, we got together as a class to share what we had found out. I asked different groups to tell each other what they had made. Here are some of the comments from the children and their explanations as to why they thought their bridges were strong.

James: I made three arches underneath . . . they were straight up and down arches, like a triangular prism. It held seventy-five Lego bits.

Christine and Rose: We made an arch underneath our flat bridge. It held a hundred little pieces *(I'm not completely convinced of the accuracy of this figure: I suspect it's another way of saying 'a lot'!)*, but without the arch it only held twenty.

Jivan: We folded it up and we had this arch. It was strong.

Me: Why do you think it was strong?

Jivan: The arch . . . 'cos underneath the paper gets thick and it's kind of forcing up the top where it stays up . . . it's pushing the paper. It's um, stapling the bridge.

Me: Stabilising?

Jivan: Yes, supporting something up.

Adam: I putted a bend in it under the straight bit and I putted two blocks on the side and I putted blocks on it, and it held it up.

Me: What do you think the bricks on the side do to the bendy bit?

Simon: The square, long blocks were sort of pushing them together so they make it more bendy.

Me: Lots of you found arches a good way to make your bridges strong. Can anyone tell me why they think an arch helped?

Annabel: *(slowly as if thinking it out and using her hands to show what was happening)* Well when you put bricks on top they push, force down . . .

Adam: Yes, and the bendy bit is pushing it back up.

Several children joined in and, as they took it in turn to describe what they thought was happening, whatever the words they used, their hands would express the pushing down of the weight on the bridge and the forcing up of the arch underneath. A few happily used the word 'force' in the sentence, picking it up from Jivan's initial introduction of the term. I asked if they could explain what this word meant. The explanations I received all revolved round the idea of one thing pushing against the other, with frequent use of hand movements to clarify what was being said.

We finished the activity by looking again at photographs of some real bridges to compare the different structures. The children could now pick out the different construction approaches and classify what they saw into three main categories: beam bridges, which they related to their folded bridges, arch structures and finally suspension bridges, which they had not attempted to make. They were also interested to see that some bridges used a combination of methods.

I was very pleased with the way the project had progressed and interested to see how much the children's appreciation of the gravitational forces involved had developed from their initial understanding. When we visited our local church two weeks later for another aspect of the 'Buildings' topic, I was also gratified to find out how well they had retained this knowledge and could apply it in other situations. The children had been to the church several times before, but this time they seemed to notice the arches that formed the Nave for the first time. James's comment on seeing them was typical: 'Look, it's got arches to make it strong'.

Lindsey's story is one that many of us will identify with, especially her delight when a child translates information learned in one context to another one, as James did. Organising her classroom so that it has an 'investigation corner' is an excellent way of keeping science on the agenda for children's learning. It also makes a strong statement that science is something that anyone can do.

—11— Exploring telephones
Jan Elliot

In this chapter, Jan describes how she set about organising and teaching a unit on 'Telephones', an item of technology that is not only familiar to most children, but of great interest to them from an early age. The children's questions focused on the different components of the telephone, but one interesting feature of their ideas was their familiarity with electricity – possibly, Jan suggests, from a previous piece of work. Jan also explains how she involved one of the parents in the teaching unit as an additional source of information. Like Rosemary Stickland in Chapter 6, Jan demonstrates how she gets her children to ask investigative questions.

Age	6–7
Situation	teachers with whole groups, small groups and individuals
Science	emphasis on children's intuitive understandings; questioning; exploration and investigation to introduce a scientist's view
Themes	to develop some understanding of how telephones work

Introduction

A telephone had been given to my classroom for our tinkering table. The tinkering table contained a lot of interesting household items, such as a radio, a calculator and an old torch. The children used this space to examine the items closely, both externally and internally. I had some tools close at hand for them to use.

The telephone drew a lot of interest from both the girls and the boys in my class. Most of the children were able to tell me something about telephones – either home experiences or from outside of the home. Telephones were clearly within their experience and of great interest to them. As a result, I decided to organise a unit of science around the topic of telephones.

Preparation

To prepare for this topic, I had to do some research. I did not know exactly how the telephone worked, and suspected that the children would want to find out things about telephones that I did not know the answer to. So I wrote away to a telephone company for information and looked up several books on the subject. From a recycling depot, I bought six dial telephones cheaply.

Materials

Overall, this unit required books and articles about telephones, six dial telephones and the necessary implements to dismantle them – screwdrivers, pliers, etc.

Getting started

In order to find out what the children's prior knowledge and beliefs about telephones were, I asked them to draw a simple concept map. The children had already had experience with producing concept maps, and were easily able to draw and write their ideas. Some children even used a mixture of pictures and writing to express their views. Where pictures were produced, I asked the children to talk about the picture, and their ideas were then written on their work. Figure 11.1 shows one of the concept maps produced.

Figure 11.1 Thomas's concept map of his understandings of 'telephones'.

The children came up with a whole range of ideas about telephones.

- They have wires under the hearing part so that electricity can go to the other part.
- When people ring up, the power goes through the cord.
- They have wires and electricity.
- People ring up and the power goes through it.
- They have colours.
- They are plastic and wire and steel.
- They can ring loudly.

- I use the telephone to ring my grandfather.
- They can ring day or night.
- Telephones can be in cars.
- They have handles.
- They have numbers.
- They have buttons or dials.
- You call the police if something bad happens.
- They have 'a to z' wires and an alligator clip clips onto cuboids.
- They have a receiver.
- When you are not home, the phone still rings.
- You can talk to someone.
- They have receivers and microphones.
- It comes from Telecom Tower and then to someone.

I found it interesting to note the ideas the children held about the wires and electricity in the telephones. It is probable that some of these ideas were the result of a previous unit on torches that I had done with them. In that unit, the children made many discoveries about electricity and circuits.

Explorations

To start with, I encouraged the children to play with the six telephones that I had acquired. Whilst the children handled the phones, they did not dismantle them. During this stage, my aim was to clarify the children's ideas on the topic and facilitate the development of investigative questions.

Encouraging children to ask investigative questions

Whilst the children were handling the telephones, I asked them what they would like to know about telephones. Some of the children needed help in asking questions, while others expressed questions quite freely. The children recorded their questions in their science thinking books. Figure 11.2 shows Samantha's recorded questions.

The questions the children wished to investigate were written on charts and displayed around the room. They were as follows:

What does it look like, the a to z part?
Why doesn't it work without electricity?
How does the bell work?
How does the wire go so far?
Why do you have to put the handle down?

Figure 11.2 Samantha's questions about telephones

Why does the voice go from you to the other person?
What are the little white things on the top?
Why does it have to go to Telecom Tower?
Does a telephone have to have a battery?
How do you talk to other people?
How do they have voices?
How does the electricity get to all the places to make the telephone ring?
How does the electricity get in?
How does it plug into the wall?
Why do you have a cord on your phone?
Where does the sound come from?
What makes it go?
Are they in aeroplanes and trains?
How does the noise come?
How can telephones be automatic?
How can they ring loudly?
How do you speak when someone is on the other side of Australia?
Why do you have to dial or press buttons to ring people you want?
Why do they need metal and wires to make them work?
How do the switches work?

I was amazed at the vast range of questions the children came up with. I think they asked much more sophisticated questions than I would have thought of, and indeed investigating these would take them much further than I would probably have planned for them. I did direct some questions that were 'safety questions'; for example, one child wrote, 'Can they plug into a power point'. The issue was immediately discussed with the class.

Once I had recorded the questions onto charts, we talked about each and thought about ways in which we could answer our questions. The children decided that taking the telephones apart, looking at books and asking experts would be appropriate methods of investigating.

Specific investigations

The children set out to work either individually or in small, selected groups. I introduced specific resources to the children to focus their thinking and help group investigation on certain aspects of the topic. For example, I read relevant information to the children from a non-fiction book, and I encouraged some of the children to explain their discovery to the whole group, particularly where I thought their explanations could be helpful in furthering the investigations of the others.

During the investigation phase, I moved from group to group, discussing findings and methods, making suggestions, questioning, and noting the progress the children were making. These sessions always ended with a sharing time, during which the children shared their thinking and findings with the whole class. Exceptions to this format occurred when I introduced slides and videos, and when an 'expert' visited the class to talk about telephones.

I asked a physicist from the local university, who was the father of a child in my class, to come and help the children with their investigations. He proved to the children that their idea about sound being converted into electrical impulses was right. He set up a light bulb in a circuit with a microphone, and the children could see the bulb light up whenever they made a noise; he set up an oscilloscope for them to 'see sound'; he demonstrated the effect of sound on a diaphragm; and he gave the children some experience of a good-sized electromagnet behaving the way it does in a telephone when the bell rings.

The children's investigations centred on the following aspects of the telephone:

- the bell;
- the receiver;
- the dial;
- the tone control;
- the inventor;
- the sound.

The bell

This part of the mechanism was isolated, and one child who had prior experience with an electromagnet immediately recognised the one in the bell system. After several children had developed theories about how the bell rang, a group was given a battery because they had hypothesised that the bell needed power to make it work. They soon discovered that they could make the bell ring by connecting the wires to the battery. They went on to discover that they had to continually reverse the wires on the positive and negative connections to make the bell hammer hit first one dome then the other.

The receiver	The children quickly confirmed their ideas that there were wires in the handset, and they made discoveries about the diaphragm, microphones, carbon granules and electromagnets. They investigated the conversion of sound to electricity and vice versa. Some children investigated the handset mechanism and discovered that lifting and replacing the receiver caused different connections to be made and broken. They theorised that the power was being cut off and connected.
The dial	Figure 11.3 shows a diagram produced by one child who investigated the dialling system. They discovered the mechanism which registered which number was being dialled and how this information was turned into an electrical signal.
The switch or dial to control the volume of the tone	The children discovered that the action of the switch or dial restricted or increased the movement of the hammer which hit the domes of the bell system, thus increasing or decreasing the volume.
The inventor	One child worked on this aspect and made discoveries about Morse, Bell and Marconi.
Sound	Some children found out about sound travelling through the air and being changed into radio waves for transmission, including via satellite.

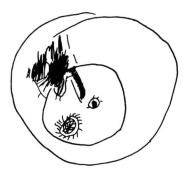

Figure 11.3   Sally's dialling system

Reflections

Towards the end of the unit on telephones, I encouraged the children to reflect upon what they knew. I gave them the choice of doing a simple concept map or just writing about what they had learned from their investigations. They were encouraged to use pictures and/or diagrams to help express their knowledge. One child recorded:

> In the electromagnet, the power goes around and around to make it ring. There is an electromagnet in the receiver. The thing what jumps and dirt flies *(I inserted the words 'carbon flies')* is called a diaphragm. It doesn't go if the power isn't on.

The children's knowledge at this stage was recorded on a large chart entitled 'What we know about telephones now'. A large number of ideas were generated, including information on the:

- wiring;
- electricity;
- vibration;
- sound;
- conversion of sound to electrical impulse and back;
- button and dial systems;
- invention;
- electromagnets;
- materials used to make the telephone.

Some of the ideas were as follows:

> If you have a battery, you can swap the black and yellow wires and the bell will ring.
> Tina

> When you talk into the telephone, it vibrates. Gemma

> There are ones where you have to push buttons. That sends a message telling which number you want. If you bump the white thing down, you get cut off. If you put the receiver down, the white thing touches inside. Then the electricity stops.
> Michael

> It was made in 1876 by Alexander Graham Bell, and the first person who invented it in 1830 was Morse. He made sound go along wires.
> Bret

> The electromagnet isn't just a magnet. The electricity has to go through the coil to make a magnet.   Sarah

> The wires need to be metal because electricity can go along metal. The bell needs to be metal because if you banged them when they weren't metal, they wouldn't make much sound.   Sophie

> When I speak into the telephone, the sound gets changed into electricity.   Nicky

The children were able to compare this final knowledge with the initial ideas which were still on display. They could see that there had been an obvious increase in the quantity and quality of their knowledge.

It was clear that the children had gained a great deal from investigating telephones. Moreover, as happens with the interactive approach, I had gained a great deal from the children. I had learnt more about what they were interested in, the range and complexity of questions they could generate, and their ability to take charge of their own learning through actively investigating answers to their questions.

In this chapter, the importance of finding out what the children already know is stressed, and the ways in which the teacher engaged the children in dialogue show the crucial role of language in science and the importance of the role of adults working with the children. It shows that even young children can explore complex technology and learn from their experience.

12 — Science from a building site
Sue Atkinson

We have included this chapter as it demonstrates how one teacher uses brainstorming and classroom discussions to find out the children's questions. It includes some open-ended questions that get children to talk and explain their ideas. The focus is on children devising their own investigations, and this is an example of teaching science with a process approach (see Chapter 1). It also shows how science can be generated from an everyday context and how motivating that is to the children.

Age	7–9 (a combined class due to the organisation of the school, with a child staying with one teacher for two or three years)
Situation	teacher working with whole class and small groups
Science	buildings; structures; materials
Themes	planning topics; starting from children's questions; children planning investigations; teacher interventions

Introduction

I had done a number of historical, geographical and mathematical topics with my class, all of which had significant scientific content; however, I knew that I had to concentrate on science for a term. Unless I did that I found that I was constantly rushing through the science and not giving the children the time they needed to think carefully about what they were doing. Even when the children had done some really good scientific work within other topics and when we had had great class discussions sitting together on the carpet, I was always left unsure as to what each individual child had learnt and what ideas they were left with. It seemed that doing science in the classroom could lead to quite bizarre ideas being formed. The often quoted proverb 'I see and I remember, I do and I understand' seems to hold true for maths, but be devastatingly untrue for science! It's more like 'I do and I get completely confused'! Observed results are sometimes the grounds for weird theories!

I'm not saying that children shouldn't do science – of course not. What I was beginning to see, though, was that, unless I took real quality time over science (and reading and discussion and maths and written work, etc.), I was in danger of thinking that I was covering the science curriculum, but actually not gaining much insight into how the children's scientific thinking was developing.

Planning the topic

I wanted my class to learn to think as scientists. By that I mean that they should use what they know and apply it to their investigations and book research. They should develop questions and be able to think and plan how they might be able to go about answering these questions.

Implicit in all of this was my belief that they wanted to ask questions and that I needed to provide an environment in which they were able to use their childhood curiosity to explore and understand their world. It followed that it was important to explore what the children themselves wanted to find out about. This was my starting point in any area of the curriculum, and although there were the constraints of a curriculum to be followed, I thought that it was still possible to explore the children's wishes and needs before embarking on a topic. This is what I did.

1 At the end of term, we all reviewed what they had learnt and thought about, that term and what they wanted to find out next term. I asked:

- what they liked doing at home (because linking home and school learning seemed to be crucial);
- what they liked in 'finding out' books (because I wanted to start with the children's interests);
- what their hobbies were and what they thought their hobbies might be when they were older (because this helps children to develop new interests and share what they know).

As I did this before any topic was decided upon, I developed quite a good profile of the children as searchers for understanding about their world over the two years that I had them.

2 We brainstormed some of the ideas.

3 We used those brainstorming sessions as a basis for future class topics and the ideas were often used by individuals or groups for their own topic work.

4 We somehow (usually with great difficulty because interests are so wide and so varied) decided on a topic. (I needed to find a balance between covering the curriculum and giving the children some freedom to choose.)

5 We then added to our brainstorm of the topic again, and the outcome was put on display (it looked a bit tatty because the writing was done hurriedly and by many different people, but as the children added to it, I think it helped them to focus on questions they wanted to investigate).

6 We then did a search for relevant books.

I found that if I did all this with the children the term before we actually started the project, I had the time to plan and prepare. I often needed to find out more about the content of any topic.

Preparing for work

One term we decided to do a topic on the building site of new houses just opposite the school. I found the site foreman, and we discussed safety (he had some spare hard hats and could get some more). He told me where I could get some copies of the site plans and house details. I told him what I was interested in the children seeing, and we planned how I could get permission from the owners to bring the children to look around. We could actually see quite a lot from the school grounds, and that would be useful in observing the stages of house building, but it would also be useful to be shown around and to see detail such as the laying of water and sewerage pipes and how gas and electricity gets to houses.

I contacted the head of science at the local secondary school, as he was always a good source of equipment we didn't have (such as spring balances calibrated in newtons, and thermometers). Most primary school science can be done with junk and bits and pieces, but sometimes more specialised equipment is needed. As ever, he was generous and supportive. I also rang up the county science centre, and they gave me books and lots of moral support; then, in the holidays, I set about reading the books I had gathered.

My aims for the topic

Although the content of the science mattered, I wanted to focus on the children planning and carrying out their own investigations. It seemed to me that this is the part of science that reveals how the children think and it encourages them to develop their own ideas. That is also one role of mathematical investigation and problem solving, and I took this 'think for yourselves' way of working as basic to my classroom organisation in every area of the curriculum. I was trying to build on what the children already know and I used discussion to explore that and develop their thinking skills. In these discussions I try to be less of a 'leader' and more of an 'enabler'. That is not an easy role but I believe it is crucial that we learn to stand back in order to empower the children.

My aims for my own professional development were to examine my interventions and also to do some careful observations of children to assess their ways of working as scientists. It was all too easy to ask the wrong type of questions or to say too much at the wrong time or to do the thinking for the child.

What we did

On the first day of term, we went to the site to have a preliminary look and to pick up some of the scraps of building materials lying around. (I had been given permission for this.) The children took clipboards to do some observational drawings, and I told them that I was expecting them to add to their list of questions about buildings that was on the wall from our brainstorming.

There were a great many questions by the end of the morning. Here are some of them.

How does the water get into the tap?
How does the man who puts the brick up know how to do it?
Why must you wear a hat?
How do the builders know where to put the house?
How much do the houses cost?
What are bricks made of?
How do you make the grey stuff that sticks the bricks?
Where do the diggers come from?
How does the electricity get into the pipes?
How do they fix the roof on?

There were so many questions that this was clearly the moment to get the children into 'interest groups' and ask them to focus on one area of their choice at a time. (I find that if children feel they have some choice, it seems to improve motivation.)

The scrap bits that we had brought back from the site were absolutely fascinating, and these were the focus of the work at first. We made a big display of them along one edge of the classroom where the children could pick things up and investigate them. There was a backing board behind this; I made this into a huge display called 'the building site' and invited children to put anything on the board that they had done or that interested them. The investigation of the scrap materials led to a great many different ideas and questions; three of these 'scraps' were breeze blocks, bricks and insulating material.

Investigating breeze blocks

One group was particularly interested in breeze blocks (the huge, grey-coloured bricks that builders use as the inside layer to house walls). The foreman had let us bring back one whole one, and in the process of finding out about it, Alice discovered that it floated. This was a great surprise to everyone, including me.

I knew from my own physics lessons that the block therefore had a specific gravity less than water – something to do with density, I thought vaguely – but how on earth could I put that over to the children? I thought I probably couldn't! (Editor note: Density is a complex concept because it is a way of comparing the weight of a given size and even fourteen-year-olds find this difficult.)

This group was engaged in finding out why it floated.

Chris: It's light.

Daniel: No it isn't, it's heavy.

Alice: Not very heavy.

Chris: Too big . . .

(after some time)

Daniel: Big boats float.

Alice: But bricks sink.

(after some time and some preliminary investigations with other bits from the site)

Alice: (using the balance) It weighs not very much. *(I think she was comparing it to the other bricks)*

Daniel: No, it weighs a lot – it's more than that stone, and that sinks.

Alice: . . . but . . . but not much. *(Was she getting at the idea that the block was lighter than you might expect for its size?)*

I was taping this session to try to look at the children's questioning and my interventions. I encouraged Alice's group to go on thinking about which other things float, and they had a scavenger hunt around the classroom to find things that could be put in water to add to their collection from the site. At first, this just involved dropping things into the water bath (they were working outside), so I suggested that they start by making guesses of what they thought might sink or float. They ended up with three groups; 'will float', 'will sink' and 'don't know' or 'it might if you are careful'. Alison joined in the work and said that in one of her books, there was a trick that showed you how to make a needle float, and some things might float or sink depending on how carefully you put them in the water. Here was the beginnings of an idea about a 'fair test', something that I had worked on with the children in previous terms.

I let them do their investigation and then asked them at review time if they thought they had done a fair test. Most thought they had, but Alison was doubtful. 'Sometimes Chris just chucked things in.' I asked her to expand on that and she explained how, if she put the jar lid in the water sideways, it sank, but if you put it carefully, 'like a boat', it floated.

Working with bricks

We had brought back a variety of bricks from the site, and Carole and some others wanted to know what they were made of. Grinding up a brick to try to find this out didn't come to much; they got better results by looking in the encyclopaedia. One group found some information about the ancient Egyptians making bricks. The children's eyes lit up at the thought of mixing earth and clay with straw. They took a few days to organise themselves and, once the bricks were made, they needed a lot of help in deciding how to test the bricks to see if they were strong. (Stacey's gran had seen the pyramids, and the idea had got around that those bricks were 'strong', unlike their crumbling lumps drying on the window sill.)

In the end, I had to intervene quite heavily and suggest that they think of a question that they wanted to know the answer to. Vikki said, 'what makes them stick together?' (She meant the brick holding together, not cement.) Katie said, 'which is the strongest brick we made?'. Others, at review times on the carpet, suggested that they put more straw, or more clay, or more water in some bricks. So, in the end, and with a great deal of input from me, another group made six bricks inside cardboard moulds (all the same size, but with differing quantities of ingredients) and let them dry out for two weeks. (One of the weaknesses had been rushing to test them before they were dry.) Then the original brick group and this second group had an interesting session, arguing fiercely about how to test them! This got very heated, but the ideas were really good, so I stopped everyone working – a thing I do very rarely – and we had a class discussion. The ideas were:

- tie the bricks onto a piece of string and swing them at the wall and see which ones break;

- hang a 'cradle' of weights over the bricks (this was a well-tried test that others had used to test the strength of plastic bags a previous term, but proved difficult in this situation because they couldn't make a strong enough 'cradle');

- drop them first from a low level, then a bit higher, then a bit higher and 'see which one is the winner';

- stack all the big kilogram weights on the bricks to see which one 'lasts the longest'.

The 'yellow fluffy stuff'

The third thing that fascinated several children was the 'yellow fluffy stuff'. This was the insulation material that the builders put between the inner wall of breeze blocks and the outer layer of bricks. Having insulated my loft and itched for days, I had brought rubber gloves to school to handle this glass-fibre and had to be strict in enforcing their use.

At first, the questions centred around 'how is the yellow fluffy stuff made?' This was one example of a 'dead end', in that no book gave us clues, but I was able to pick up Wendy's question, 'does it really keep the house warm?' and I asked her how we could find that out. This group had various grand plans about sending questionnaires home to ask about how warm houses were, but in the end I suggested that they think about investigations they could do in class, trying to keep something else warm, rather than thinking just about keeping houses warm. (At first, I wasn't really convinced that the children would be able to make that leap to a different context, but some of them did and these investigations were a source of considerable learning.)

The work with this group was hard going at first, and I had to intervene quite heavily to get them thinking of what they could do. It was a cold day, and with the class, I discussed how we kept warm with our coats. They picked up on this, and discussion ranged around:

- coats;
- animals with fur;
- birds puffing up their feathers;
- thermos flasks.

They discussed how they could test out whether the 'yellow fluffy stuff' really kept things warm, and this proved to be a good area for devising investigations, because it was fairly easy to change some aspect of it.

One investigation was to put different amounts of hot water in yoghurt pots and wrap each pot in the equal amount of glass fibre, using the thermometers to take the temperature every five minutes. Another group joined in the work, but wrapped the pots in various coats to find which coat kept in heat the best.

After a couple of weeks, we were well into the project, and each day we were discussing:

- their hypotheses;
- whether the investigation was fair;
- whether there were too many different factors involved to be sure what we observed would happen again;
- whether it answered the question being asked;
- whether what was predicted actually happened;
- what to do next.

On the building site display board the children had put up their questions and what they had found out. I was impressed with the way in which they were working, and thought how their independence was growing. They frequently asked me for advice and facts (and to settle disputes), and I found that the way I felt most comfortable handling this was not to tell them facts – I often didn't know the answers to their questions anyway – but to encourage the idea of finding out together. That is not to say that I didn't ever tell them facts or share what I thought I knew, but I didn't want to become the type of teacher that children perceive as having all the answers.

What sense does the child make of what they see?

Making sense of investigations seems to be the meeting point of facts, the child's previous understandings, observed phenomena that we can't always explain, and this crucial and difficult area to explore – what sense the child puts on what they see.

It was this area that I was trying to explore in our whole-group review sessions. What sense was Alice making of the floating block? What did Kerry and Jenny gain from grinding up the brick? Were the children really able to relate those investigations with hot water and coats to houses and insulation? I'm not sure that I ever found the answers. What I *did* find was that, by encouraging the children to record their work for themselves and then talking to them, their ideas and questioning became really quite sophisticated by the end of term. The question about the water getting into the taps developed into a linking between the siphon that we had worked on during a previous term and how the water gets up to the tank in the roof. The issue of water pressure was discussed, and I found my own scientific understandings pushed to the limit. (I now use my own children's science text books from secondary school, as these put over complex ideas in a simple way.)

The importance of talking

It seemed to me that it was the process of talking that clarified much of what was going on. The 'doing' seemed to lead to all sorts of confusion. (Kelly thought that we were making the water in the yoghurt pots hotter by wrapping them in coats.) Children writing and drawing about their work seemed painfully slow compared with the rapid exchanges of ideas in discussions and in review times, but the value of that time spent in written work paid off in that it gave me time to talk to individuals about what they thought and the development of their ideas.

I found that, as I listened to myself on my audio tapes, some interventions were better than others at getting the children thinking. The ones that helped the children were the questions that also helped in maths – and in geography, history and sometimes in language.

'Tell me what you are doing.'
'What are you trying to find out?'
'What do you think might happen?'
'Why did you think that was so important?'
'Do you have a question that you are trying to answer?'
'Tell me why you think that.'
'Can you say a bit more about that?'
'What did you find out?'
'Did you manage to answer your question?'
'Which other things might be important?'
'Do you think that will always happen?'
'What if you . . . ?'
'Why do you think that is?'
'If you did it again, what would you do differently?'

What had they learnt?

At the end of the project, I asked the children what they thought they had learnt. They said a great many different things, which was very reassuring. Among them were many areas of scientific knowledge:

- 'how to build a house';
- 'what goes underground';
- 'why the roof slopes';
- 'what things are made of';
- 'how to keep things warm.'

There were other areas of scientific thinking:

- 'is it fair?';
- 'how to find things out';
- 'measuring it carefully'.

I thought, in addition to this, that they had learnt a great deal about:

- group co-operation;
- how to think about a problem and suggest ways to find answers;
- how to observe closely and carefully;
- how to build in their previous knowledge;
- how to listen to others;
- how to integrate their book knowledge with what they observed;
- how to think in a logical way;
- how to form hypotheses and test them out ('this brick will be stronger than this one because . . .', 'the water in this jar will stay hotter because it has more insulating stuff round it', 'this wall will be stronger than that one because it has more bricks in it').

The class big book on 'the building site' worked really well, and we could have made it twice the size and still have filled it. The interactive display of the bits and pieces from the site was a source of fascination all term. At the end of term, they wanted to keep all the things and not get rid of them. Several things went home to add to collections of precious things. It was interesting how quite ordinary things, like a piece of plastic drainpipe, stimulated conversation and questions that lasted for weeks.

The children's individual books for the term were obviously mainly scientific, but they also included:

- stories;
- poems;
- a great deal of maths (shapes that fit together, scale, measuring of angle, temperature, area, volume, linear measuring);
- historical information from buildings in the centre of town and from nearby villages (date stones on the front of houses, etc.);

- observational drawings;
- maps and plans;
- the children's own designs of future houses, which they had great fun with and I thought demonstrated great ingenuity and creativity (Paul designed an underground house, Amie had solar panels on hers, and Chris's one was based around a train set that went from room to room and brought him his breakfast in bed!);
- various printouts from the computer database.

What I learnt

Following what the children were interested in helped me to feel more confident about teaching science. I could see more clearly how I might assess it, because focusing on the investigations that the children devised for themselves gave me considerable insight into the children's scientific thinking. I learnt a great deal about my own interventions (I found listening to tapes of myself excruciating sometimes!). I was often inappropriately 'closed' when discussing things with the children and often did not really 'hear' what a child was saying until I listened to the tape.

When I reported to the parents at the end of term, I could say a great deal about each child, and many parents expressed their surprise and pleasure at what their child had done. I had found the project enormously exciting and enjoyable, but I was still left with significant questions, such as was I really dealing with scientific facts adequately? I wondered if I would ever resolve that.

How do we teach science? What materials do we need? What will the children learn? This chapter hopefully gives some answers to these questions and shows that there is a wealth of work that can evolve from investigating the world around the school. Here we have seen children generating their own questions for investigation and beginning to talk the language of science.

13 — Exploring outer space
Karina Sargeson

We have included this chapter because outer space is something that is of special interest to children, no matter what age. However, many of the concepts that need to be taught under this topic are not only challenging to teach, since they cannot be concretely manipulated, but also difficult for children to understand. For example, conceptualising that the world is round and not flat is difficult even for adults, as a quick look at history will show! The rotation of the earth in relation to the sun causing day and night is also confusing (especially as we talk about the 'sun set', the 'sun rise' and watching the 'sun come up'!). Our everyday language does not assist children make sense of these complex ideas. Karina describes how the children's concept maps clearly illustrate the complexity of their thinking, as well as the weird and wonderful ideas they hold at the age of eight. The experiences Karina provided to make the unit more meaningful are presented alongside some of the children's work.

Age	8
Situation	teacher with whole groups and small groups
Science	investigating children's understanding of outer space
Theme	an integrated approach to teaching science, using videos and printed materials to investigate an area which children cannot physically explore

Introduction

Developing knowledge and understandings of the earth and its place in space is both important and exciting for children. The potential content to be covered is seemingly endless, and the children are amazed and excited by it. The topic of space is also one that can be readily integrated into other curriculum areas, for example creative writing, movement and dance.

Comments and questions about space came up in my class frequently, and the children often have interesting pieces of information to share about the topic. In free-choice activity time, several children chose to research different topics related to space.

Due to the interest amongst the children, I decided to study the topic in depth during the second term as part of my science and language programme. The approach I took to teaching this unit was to begin with what the children already knew, generate questions of interest to answer individually and as a class, do activities together to build a joint field of knowledge and encourage the children to do research and experiments individually and in groups.

Over the term, it became evident that the children were taking in and retaining a vast amount of knowledge. They were enthusiastic about sharing their understandings, and their interest in the topic was maintained over the term. In fact, some of the children would have been happy to study space every day for the rest of the year!

What I did

The unit progressed through the following specific topics:

- our solar system;
- the planets;
- space exploration;
- stars and nebulae;
- astronauts;
- space vehicles.

Each of these topics was covered in a variety of ways:

- language activities;
- shared reading;
- maths activities;
- art work;
- excursions;
- class research;
- videos;
- posters;
- co-operative group work;
- research projects;
- technology activities.

The approach to this unit was integrated into the overall curriculum. The children had many opportunities to work independently or with peers on their own projects and research. The emphasis was on the joint construction of knowledge as a class and as individuals. Our shared knowledge grew extensively and each child also gained knowledge of particular areas that interested him or her.

The expression of this knowledge was again through a variety of media:

- big books;
- posters;
- display boards;
- factual reports;
- constructions;
- art work;
- projects;
- models.

The result was that all available space was taken up with space creations and pictures! The children certainly had the satisfaction and pleasure of seeing their work around the room, and the constant presence of work we had done together made it easier to refer to and remember.

The children's work

The following are examples of the children's work throughout the unit; I have tried to select work that reflects the general ability of the group. What cannot, of course, be included are the class discussions, group conversations and teacher–child interactions that occurred daily throughout the term.

Individual concept maps

Figure 13.1 Child's concept map.

These were the starting point; they enabled the children to begin their own thinking about the topic. They also gave us, as a class, a common starting point that established what we already knew as a group. At the conclusion of the unit, the concept maps were an indication to the children (and myself) of just how much they had learnt. Figure 13.1 shows an example of one of the concept maps.

Question time

After sharing our existing knowledge, we generated questions concerning specific things we wanted to find out about space. The children shared their questions. Each one was recorded in a big book entitled 'Questions from out of this world'! Over the term, the answers were recorded in the book.

Big book – 'Earth and the planets beyond'

Our first information-giving session was to watch a very informative video about the planets in our solar system. We watched it twice (on different occasions). The second time through, we took notes about each of the planets. This gave the children real incentive to pay close attention and to focus on the main points mentioned in the video. After taking notes and discussing the contents of the video, we wrote a big book together entitled 'Earth and the planets beyond'. We revised the characteristics of a factual report, collated our information and produced a text full of fascinating facts about each of the planets.

Fact cards

The children were so excited about their new knowledge of the planets that a group of them decided to make a model of the solar system. They wrote a fact card for each planet with information about temperature, size and distance from the sun.

One of the maths lessons involved the children in representing the solar system with planet size and distancing to scale. Whilst scale was a difficult concept for the children to grasp, they certainly understood the vastness of the solar system and the incredible differences in the sizes of the planets. For example if the earth was a tennis ball and the earth's moon a marble, then the sun must be as big as the classroom!

This maths activity gave the children sufficient knowledge to make paintings of the planets with reasonably comparative sizes and to space them along the rafters from which they hung. This model and the fact cards were referred to often throughout the unit.

Excursion to a space tracking station

To provide the children with a base of knowledge from which to spring into other areas, to impress them with the vastness of this topic and to show them that researching space and working in that field are very real and exciting possibilities, we went to a space tracking station. This proved to be an invaluable excursion. The education centre was filled with models, displays, photographs and information that the children could easily understand and record. Seeing photos and videos of the astronauts who have been into space really excited the children and made it all seem very real. The children acquired a great deal of knowledge from this centre, and their interest and excitement rose even further.

Over the next few days we shared the information gained, and many children made models of the shuttles, tracking dishes and various things they had seen at the space tracking station. The children would then find out some more information and recall details from their visit to write an explanation about their construction.

Art work

After watching a programme about stars and nebulae in outer space, the children painted their own nebulae on black cardboard. The effect was quite dramatic. The children learnt that nebulae could be in many forms and colours, so their paintings reflected this knowledge. The children were blown away by the vastness of space and the number of stars in our solar system!

Acrostic poems and alien mail

One of our language lessons involved the children writing acrostic poems about space and also poems or rhymes to help them remember the names of the planets. For these poems, the children listed the initial letters of all the planets in a line and then made up a sentence to help them remember the order of the planets. The children had some very creative ideas, although the challenges for them was to ensure that the rhyme made sense rather than being isolated ideas (Figures 13.2 and 13.3).

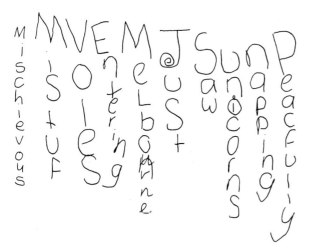

Figure 13.2   Example of a space poem. It reads, 'Mischievous voices entering Melbourne just saw unicorns napping peacefully'.

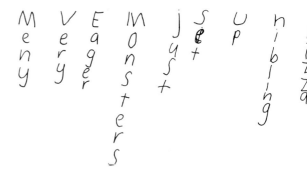

Figure 13.3   Example of a space poem. It reads, 'Many very eager monsters just sit up nibbling pizza'.

Space food

One particular topic the children found especially fascinating was space food and how the astronauts actually ate in space. Once again, we watched a terrific video from which the children took their own notes. After listing the great variety of foods that astronauts ate, the children, in groups of four, wrote a menu of the foods they would choose for a day if they were in space. This became a class menu for an imaginary journey into space (Figure 13.4).

Space suits and space crafts

To be able to travel into space on our imaginary journey, we needed to find out about space suits and space craft. After sharing some

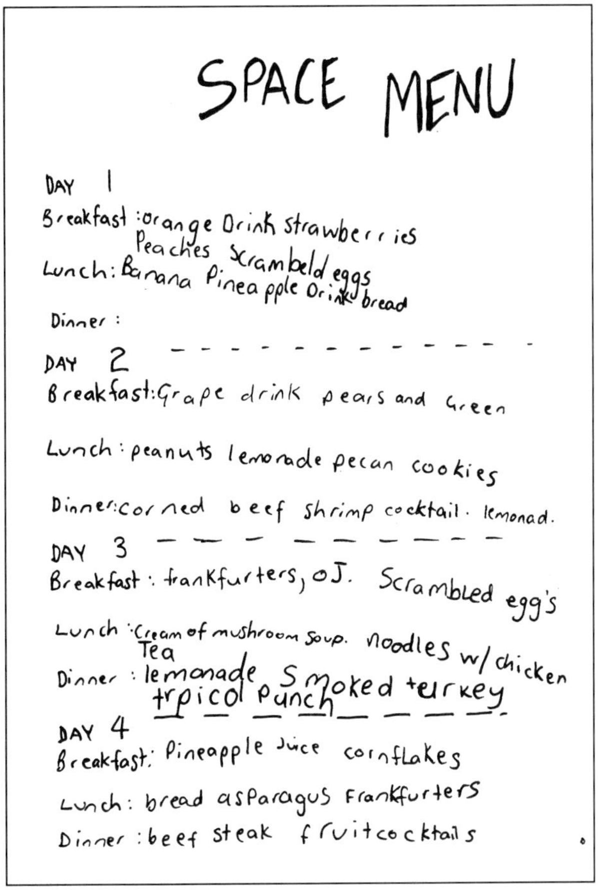

Figure 13.4 Space menu

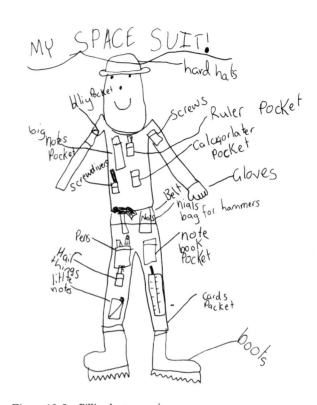

Figure 13.5 Jillian's space suit

information and discussing helpful pictures, the children designed their own space suits, with pockets and pouches for all the things they might need to carry on them in space (Figure 13.5).

To design our space craft, we watched a video showing different space craft and researched some information. The children drew diagrams of the interior of a space craft. Some children also did their own research and designed new types of space craft or modelled space craft from illustrations in books.

Independent projects

As part of our daily work, the children usually have time to plan and engage in a great variety of their own activities. Many of the children frequently chose to do projects related to space. Some of these projects were:

- technology, involving construction and design;
- researching projects and writing factual texts;
- studying pictures and producing art work;
- studying illustrations and making models;
- writing plays or stories and presenting drama.

The children gained much from the different activities they engaged in, and a highlight of the week was always our sharing time, when children had the opportunity to present to the rest of the class the projects they had done during the week. This was a valuable way both of observing what they had learnt and also for the children to articulate their learning.

Writing non-fiction books

To conclude our unit on space, the children wrote up their own, individual non-fiction books. We generated headings as a class to form the structure of the text, then individually or in pairs they wrote down all they could remember about each topic. Each of their texts was 'published' and shared with the rest of the class before being put on display in the school.

Conclusion

The excitement, enthusiasm and interest shown by these eight-year-olds throughout the entire unit made this a terrific topic to teach. The children gained a broad range of knowledge about the subject and were able to express their knowledge in a variety of ways and articulate their understandings clearly to others. They delighted in sharing every new fact they discovered with one another and their families.

It was a unit that readily integrated into other curriculum areas, making it an integrated unit of work. The emphasis on sharing and interaction meant that we built a shared field of knowledge for our work. An exciting concluding activity using this shared knowledge was writing a maths 'big book' for a competition. The focus of the book was the visit of two aliens to earth in the future. These aliens wanted to know a range of mathematical facts about earth and earthlings. The text gave information to the reader about the alien planet and asked corresponding questions about earth. The children enjoyed sharing and extending their knowledge in the writing of it prior to sending it away for the publication competition. Unfortunately it didn't win anything, but that didn't diminish the children's pride and pleasure.

The children developed a range of skills above and beyond the content matter that will remain with them and assist them in future work, particularly in language. Overall, exploring space was a rewarding unit for these children and the learning that resulted from it was outstanding.

Many of us find 'space' hard to teach because of the need to do so much work from books. Karina has captured the enthusiasm of these eight-year-olds and shown how videos and books can stimulate science learning. She shows how the topic captures children's imagination and stimulates them to produce such wonderful, creative work.

Astronomy is the oldest science and is currently a very exciting area of modern science. Where are we? How big? How old? These are all questions which this topic raises. Some of them can't be answered just by science – they are huge questions about our very existence – but they certainly fascinate children.

14 — Helping to save our planet
Karina Sargeson

In this chapter, Karina describes how her children explored the concept of energy through designing and making energy-efficient homes. Various forms of energy were considered by the children, such as wind power, solar power and water power. The children designed the exterior and a plan view of the interior, as well as providing a written explanation of how their energy-efficient home functioned. An array of interesting homes was designed by the children.

Age	8
Situation	teacher with whole groups and small groups
Science	investigating children's understanding of energy
Themes	designing and making energy-efficient homes

Introduction

The concept of energy is a difficult one for young children to understand, yet for many reasons it is important that children are taught at an early age, about energy and its uses. In fact, children can develop an understanding of energy from an early age, if the concept is taught in a simple and meaningful way.

As with many science topics, choosing the right approach is the key to helping the children develop correct understandings and scientific knowledge. Providing useful, clear and concrete materials is vital. Interacting with the children and helping them to move to a higher level of understanding is also essential. The role of the teacher on a teacher–class and one-to-one basis for interaction and discussion is perhaps the most important aspect of teaching, particularly when it comes to complex ideas.

Starting points

Our unit on energy began with class discussions and continued in this way, with constant interaction between teacher and children and children with their peers. A very important aspect of our work involved developing a shared vocabulary of words related to this science unit.

The ideas the children expressed were:

Life now and then

> 'Our way of life depends on having a constant source of power and fuel.'

Old way of life

- kerosene
- candles
- horses, carts
- wrote letters
- time from the sun
- hand washing
- baths
- fires for cooking
- hand irons heated on fires

> 'Now we use electricity and fuel for everything.'

Our ideas for new energy sources

Instead of using up oil, coal and gas, we could . . .

- use skylights;
- use wind power;
- use water power;
- walk;
- use solar energy;
- use solar heating;
- use flower power for buses (oilseed rape instead of diesel).

This is important in all science work, as there is often a language that is unique to the topic being discussed. The children increased their knowledge and felt much more confident because they were able to use the appropriate language at the appropriate times when discussing the work we were doing.

As part of our everyday work, the children were given frequent opportunities to plan their own activities and engage in them for blocks of time. A popular activity was to design and construct machinery, equipment and vehicles for the future. A couple of children were actually designing a futuristic city that they wanted to be completely non-pollutive and environmentally friendly. This led them to ask questions about how to do this. Other children, who were also making various inventions for the future, had the same need for further information.

This early interest and need for information was what led our class into a unit on energy. The children already had some knowledge of energy, what it was and what it was used for. They had sufficient knowledge to be able to ask questions to find out more. All the children understood the concept of human energy and also that energy is used to make things work. This became our starting point.

What is energy?

Our first step was to discuss what energy is. The children had varying ideas about this:

- food gives us energy;
- it can be in your body;
- it makes you feel active;
- there is solar energy;
- energy can pollute;
- it can be in batteries;
- it can be dangerous;
- there is electrical energy;
- energy is like power;
- it makes things work;
- it's in the car.

We read some information together that gave a clear idea of what energy was and together wrote what we now knew about energy. The children could easily see which of their ideas were correct and added some new ideas. We also read some fascinating facts about the current state of energy supply and the way humans have used resources so extravagantly. This rather amazed the children, and they responded in a very positive way by wanting to know what we could do to save resources. This led us into the next phase of the unit.

Saving energy

After reading about energy, we summed up with the following:

- if we had no energy, we wouldn't be able to do all the things we do;
- energy is provided by gas, fuel, electricity, food, water, sun, wind, oil and coal;
- we use these forms of energy to give us light, heat, air-conditioning, communication and electrical power.

The unit on energy was very closely linked to our work on environmental education. The children had a genuine concern for our environment and were now keen to understand ways in which we could help save our world through using existing energy sources differently or using alternative sources of energy. So began a unit of work spanning several weeks.

As simple resources were difficult to come by, I made a series of posters with written text, illustrations and diagrams that showed the alternative sources of energy and gave clear information about them. For example, I had a poster on wind power, one on solar power and one on water power. These were very useful for developing the children's knowledge and as references throughout the unit.

As a class, we discussed these posters with information about energy and, more specifically, about wind, water and solar power. Working in pairs, the children then researched one of fifteen machines (for example a wind farm and a fuel cell, shown in Figures 14.1 and 14.2) that made use of alternative power sources. Each one was simple enough for the children to understand and had clear illustrations and diagrams. The children presented their work as mini-posters. When each of these was shared with the whole class, the children had quite a pool of knowledge to draw from.

Figure 14.1 Wind Farms by Amy Vermeer

The children then launched into various projects themselves, constructing a great variety of things. Here are some examples of their comments about their constructions.

'I made a house that doesn't need to use any electricity for the lights. There is a skylight for the whole house which is really big.' Amy

'I made a country cottage. The house is run by solar power. There is a tank beside the house to collect rain water. There is a solar panel on top of the tank that collects the sun's rays, which are used to heat the water. There is also a solar panel on the roof of the house which is used for the heating system, so that there is power for cooking and power for other appliances.' Lucinda

Figure 14.2   The Fuel Cell by Daniel Kuzmins

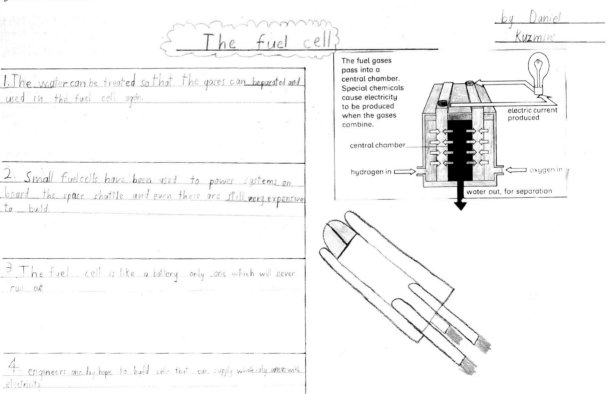

'I made a model of a solar panel. I got a silver piece of cardboard, some plastic and some foil and four pop sticks. I stuck the plastic on top of the pop sticks, then I stuck the foil on the sides to reflect the light and to store it as heat.'   Lachlan

'I think that boats in the future could use the wind power. One way would be to have the wind blowing into a propeller to make the boat travel along. Air could go into a balloon-type thing and then, as the air is released, the boat moves along.'   Nathan

'I made a paper recycling machine. On the top of it was a square solar cell that gave power to a lever that started the machine running. The solar cell absorbed the heat. The heat panel was connected to the main levers which, when hot, would run the machine.'   Anthony

'I made a desk lamp for a school desk. On the roof, there would be a solar panel that collected the sun's light and heat. It collected it for the night time and the day time. A faint cord connected the lamp to the solar power and made the lamp work.'
   Kylie

There were numerous other interesting inventions, including solar-powered rockets and robots! The children seemed to have an urgency about inventing things that always used alternative power sources as their only source of energy. Although they weren't always able to articulate exactly how the energy was accessed and put to use by their machine, the understanding of the need was there, as well as the interest in exploring the technology needed to make their inventions a reality.

Other activities

Whilst the children continued inventing things and the city of the future began to take shape and form, we launched into some other activities as a class. We began a major project that involved the children in working in groups to design, make and appraise their own house for the future.

Our first step in this project was to visit the National Science and Technology Centre. The children spent a block of time in the Environment Gallery, answering questions and doing their own research to gain ideas for designing and building their house. This was an exciting excursion and proved to be a fantastic source of information for the class.

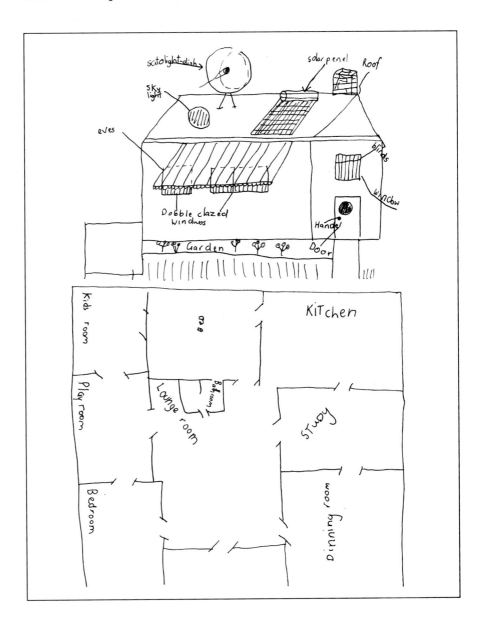

Figure 14.3 Child's design of a solar house

With a model house and a variety of displays demonstrating how energy is used, ways of reducing our consumption and alternative sources of power, the children were able to gain a lot of new ideas.

A science show on energy was also helpful for the children in demonstrating the different types of energy that exist and how they all work together or work in different ways.

Back at school, we collated our ideas in the form of charts and posters, which were displayed around the room. The children then began to design their house for the future. Their task was to design a poster with the following elements:

- house name;
- drawing of exterior;
- aerial view of interior;
- garden plan;
- everything clearly labelled;
- explanation.

The children then had to build the house that they had designed and complete the explanation, giving clear information on how everything worked and why it was an integral part of the house.

The children launched into this project with great enthusiasm. They felt sufficiently confident about their knowledge of the topic to be able to put it into practical use. They had the language they needed for the explanations. There was a great variety of clear and informative posters available for reference.

The results of this project were outstanding. Plans were clear and detailed. Ideas were realistic and scientifically correct. Models were well presented and quite complex in their making.

Overall, this was an exciting unit for the children. Their interest and enthusiasm were astounding, and they had a genuine desire to learn about the effective use of energy, both for now and the future. The knowledge these children have and the confidence that they place in their ability to research, write about and design inventions for the future is exciting. They may one day make a huge difference in our world!

Energy is a difficult topic because the scientist's concept is such an abstract one. Children (and adults) often find it easier to think of fuels and where they come from. What kinds of different fuels are there? What do fuels do for us? Exploring these questions provides a good concrete platform on which to develop their thinking later.

Karina again shows us how she captures the interest of her class with a project they genuinely enjoy. And, as she says, these children may one day contribute to saving our planet!

15 — Them bones, them bones are not all dry!
Mary Sofo

In this chapter, Mary sensitively describes a unit of work on bones that she did with a small group of children. Mary not only outlines how she organised the learning experiences following the children's interest, but provides many reflective comments on the challenge of teaching science to children. This challenge is evident in the questions the children posed: 'Do insects have bones?'; 'Are they male or female bones?'.

Age	9
Situation	teacher working with a small group of children
Science	exploring bones
Themes	preliminary investigations; exploratory activities; children's initial questions; helping children answer their questions

Introduction

Jessica started to rhythmically bang two of the bones together. I thought this might be a good opportunity to explore ideas about bones and their hollowness. Not wanting to intimidate Jessica, who was rather withdrawn and shy, I thought I would refrain from asking, 'why are you banging those bones?'. What I did ask was, 'what sound are the bones making?'. On reflection, Jessica should have had the opportunity to tell me why she was banging the bones, and then I would have proceeded to ask a more appropriate question. Instead Jessica's reply was, 'I don't know'. Vincent, who was listening to our conversation, quickly answered, 'they sound hollow'. Jessica was invited to bang the bones again.

Me: What do you think, Jessica?

Jessica: I don't know.

Vincent: They sound hollow because bones are hollow.

Me: What do you think of Vincent's idea, Jessica?

Jessica: I'm not sure.

It was obvious Jessica felt uncomfortable and was not interested in pursuing the subject. The question Jessica wanted to investigate was whether the bones were male or female, and how you could tell the difference. Vincent was not interested in pursuing the subject any further. He was more interested in asking if insects had bones.

In my eagerness to interact effectively and ask searching questions, I became a little more self-conscious and less effective than I would have liked. I felt better questioning may have helped Jessica to ask a different question.

Why study bones?

My reason for selecting this topic to explore with children were based on the following beliefs. Firstly, bones are generally fascinating to both girls and boys. They are a familiar part of their world and well within the realms of their experience. Secondly, I felt that, as a topic, bones could provide a range of rich experiences and prompt the children to ask scientific questions about them.

The children certainly asked questions. For example, during a private discussion between Erin and myself, she wanted to know why are bones so hard and whether you could break them with a hammer. After listening to Vincent's question, 'do insects have bones?', she then changed her mind and wanted to ask, 'do ants have bones?'. Discussing this point further, Erin stated that she really wanted to know if ants had bones before she pursued any other question. This was an interesting situation. My reaction was to allow Erin to satisfy her curiosity. The following description outlines how I went about helping Erin and the other five children I was working with.

Preparation for teaching

In my background reading for teaching a unit on bones, I found that there were six important points for the children to think about. They included knowing that bones:

- exist in vertebrates only;
- serve as a structural frame to support the body;
- enable movement by providing a point of attachment for muscles and by serving as a system of levers;
- protect vital organs, such as the brain, spinal cord and the soft internal organs;
- house the blood-forming system (red bone marrow);
- act as reservoirs for the mineral calcium, which is vital to many body processes.

Assembling resources

Dry bones

Since bones are so much a part of our everyday lives, finding suitable hands-on material is easy. As part of my preparation of the topic, I was fortunate to come across some human bones housed in the science cupboard at the school. This box consisted of a human skull, ribs, a hand and several leg bones. The science cupboard also contained the skulls of a horse, a cow, a sheep and a bird. There were also other animal bones of different shapes and sizes, and bones of varying ages and degrees of decomposition.

Fresh bones

I bought some fresh bones from our local butcher. I asked the butcher to saw the bones down the middle so that the children could see the marrow inside the bones. I encouraged the children to remove the marrow and thereby discover that bones are hollow. It was hoped that this experience would help raise several questions, such as, 'why are bones hollow?', 'what is marrow?', 'what is its function?'.

Magnifying glasses

I set up a corner of the room with magnifying glasses and things to measure with, such as rulers, tape measures and kitchen scales. These were used to look at bone tissue more closely, to weigh and measure bones and to help raise questions for further investigations. For example, why are bones different shapes and sizes? Why are the ends of bones spongy on the inside?

Books, charts and audio-visual materials

Other resources that I used were books, charts, models and audio-visual materials, especially a video entitled 'The Skeleton'. This video briefly discussed the evolution of the skeletal system. It provided excellent information on the growth and formation of the system that provides protection to parts of the body. This video addressed the six important aspects of bones.

Guest speakers

If one of the children's parents had been an osteopath or chiropractor (or some other profession with a knowledge of bones) we could have invited them in to talk to the children. Unfortunately, we were unable to do this, and had to rely upon books.

Excursions

Another good source of information for this topic is a museum that has fossils, dinosaur bones and other animal bones. An excursion to a museum would also be a great learning experience for the children.

Explorations

Before I began teaching, I organised a session where the children were given opportunities to play freely with a box of assorted bones. During that time, I asked them informally about their ideas on bones. I also asked them to draw what they thought the inside of a bone looked like (Figure 15.1). The children's ideas were:

Figure 15.1 Child's drawing of a bone.

- bones give things their shape and size;
- they help you move and pick things up;
- they protect your heart;
- they are made from the same substance as your teeth;
- they are smooth, hard and strong;
- they have tissue in them, a 'kind of marrow';
- they fit into sockets;
- they all do a different job;
- animals have them as well;
- they are surrounded by flesh;
- they have calcium in them;
- they are joined to other bones;
- bones are hollow 'because in your head brain is in it';
- bones are hollow 'because when you pick up dead animal bones, they are hollow inside'.

This investigation showed that what the children already knew was reasonable, fairly comprehensive and not in conflict with expert opinion (Biddulph and Osborne, 1984). Our group discussion also revealed that there was one basic assumption that the children held about bones which needed clarification. The children used the words 'bones' and 'skeleton' interchangeably and took them to mean the same thing. They assumed that, because bones are part of skeletons, then all skeletons are made of bones. Although the first part of this statement (bones are parts of skeletons) can be considered correct (the reason being that bones only exist in vertebrates), the latter part of this statement (all skeletons are made of bones) is not necessarily true. Some creatures (insects and crustaceans) have skeletons known as 'exoskeletons', but these are not made from bone, they are made from other substances. This assumption therefore raised two questions which children could have decided to investigate.

1 What are bones made of?
2 Are all skeletons made of bones?

Defining children's concepts of bones

Activity

In the light of the above assumption, the following activity was designed to help clarify what the children meant by 'bones'.

The children were shown a series of pictures and objects (instances of bone and non-bone) which they classified as bone or not bone. The objects were:

- the shell of a cicada;
- the claw of a crab;
- a cockroach;
- different bones of animals;
- a shed snake-skin;
- the outer shell of a cray fish;
- a caterpillar

This activity revealed further information on how the children defined bones. I realised that further discussion, experimentation and consultation with 'experts' (in this case, books) was needed so that the children would come to realise that not all skeletons are made of bones.

Exploratory activities

To set the children thinking about bones, some exploratory activities were organised. It was hoped also that these activities would raise some questions for the children. For example:

- Why are bones different shapes, sizes and thicknesses?
- What are the smallest/largest bones in our body?
- Where are they situated?
- How do bones move?

Activity 1: feeling for bones

I asked the children to touch different parts of their bodies, starting with the centre of their backs, and asked them various questions.

- What can you feel?
- What are those bumps?
- How many bones can you feel in your hand?
- Can you feel their shape/size/length?

- How many ways can you move them?
- Feel your chest. How many bones can you feel?
- Feel your head and face. How many bones are there?

Activity 2: no thumb?

The children were asked to tape down their thumbs into their hands. They were then asked to try and perform several different tasks, such as picking up a pencil, scratching their noses or writing on a piece of paper. A list of activities normally performed in a day was then compiled, and the children considered the following questions.

- Which of these activities rely on the use of the thumb?
- Which do not?

Results were recorded and discussed. These activities stimulated the children to think about how bones are structured and the way they are arranged, and how these factors greatly affect what we can accomplish. This also helped them to realise the importance of bones in relation to movement.

Further awareness of body movement and its relationship to bones was achieved through drama sessions.

Activity 3: bone drama

I introduced the session with a poetry reading, taking movement as the theme. For example, Eleanor Farjeon's poem 'Dancing' invites simple imagery of the many ways of moving. The poem is also one children enjoy moving to. The children were helped to stop and think about, then demonstrate, the various movements called for. These movements included and demonstrated *locomotor movements* (in which the whole body moved from one place to another) and *non-locomotor movements* (which is when a part of the body stays in one place).

Further discussion allowed the children to realise that we are able to move in many ways because of the way the body is structured, and that bones are what structures our bodies. They were also asked to imagine and enact being a body without bones.

Helping children answer their questions

Before we became too deeply involved in the topic, I wanted the children to have plenty of time to explore their questions. I felt that the children needed to be really clear about what they wanted to investigate.

Session 1

Firstly, the students were encouraged to handle freely and explore the bones. Time was given to do this without any interference from me. This provided a good opportunity for me to closely observe the children's reactions and any particular interest individual students showed towards certain aspects.

Fresh bones were also placed in a separate area, with things to measure nearby and the children were encouraged to explore these as well. However, they did not appear to be as interested in these.

The group displayed enthusiasm and excitement at the prospect of exploring the bones and needed no prompting in using the different kinds of investigative equipment (especially the magnifying glasses). However, the girls were a little hesitant and reserved at first and needed a little coaxing. Then after ample exploration and interaction with me and each other, this session ended with the children refining their questions in a definite and concise manner, vocalising them to the group with confidence and writing what they would like to investigate in the 'think books'. Their questions, which were also recorded on large sheets of paper and displayed, were:

> Do ants and insects have bones?   Erin
>
> Are the cracks on a human skull normal, or has the person had their skull cracked?   Fiona
>
> Is the human skeleton male or female and how can you tell?   Jessica
>
> How many bones are in a human body?   Denis
>
> Do insects have bones?
>   Steven and Vincent

Some of the questions that could have emerged, or that I could have introduced, are shown below.

- How do bones move?
- What makes bones strong?
- What are bones made of?
- How do bones grow?
- Why do we have bones?
- Why do bones have marrow?
- Who has bones?
- Are all skeletons made of bones?

Session 2

Our second session focused on helping the children consider what the present answers to their questions were and record them in their 'think books'. They were also encouraged to look at charts, models and books that were carefully chosen and placed around the room. These resource materials held information that addressed their questions.

During this session, the girls felt more relaxed and confident. The children seemed keen to look at the reading material provided. There appeared to be more group discussion. Steven, a little boy with reading difficulties, brought his own book on insects that helped him answer his question (Vincent and Steven were close friends). Vincent took on a peer teaching role, so both of them had decided to investigate the same question: 'do insects have bones?'. I felt this demonstrated Steven's interest and level of commitment to the topic.

Entries in their 'think books' at this stage indicated the children's use of language and the level of their thinking and knowledge on the topic. Their ideas, before serious investigation, were:

> My thoughts are that ants and insects do have bones because how else would they move? But maybe they don't have bones. Some people think they have bones on the outside like on the shell.   Erin

> I think that the skull was cracked and that he/she tripped over.   Fiona

> I think he was a male and he was about fifty to sixty years old.   Jessica

> There are over 200 bones in one person's body.   Denis

> I think insects have bones on the outside of them – it might be a type of shell.   Steven

> Insects have bones in their wings, and if they don't have wings, they have bones on the outside of their bodies.   Vincent

Session 3

In our subsequent sessions, the children continued to investigate their questions by searching through specific sections of resource materials and watching audio-visual material. Fiona, Denis and Jessica, who asked specific questions about bones, for example, 'how many bones in a human body?', were satisfied with their investigations and proceeded to formulate other, related questions. Denis, who asked the above question, wanted to know if there were more bones in the hand than in the foot.

He proceeded to sift through literature to find his answer. However, Erin, Steven and Vincent, who asked, 'do insects have bones?', needed to discuss their findings a little further, because their ideas were unclear.

Under the heading 'what I found', Erin wrote, 'what I found out was that ants don't have bones because the skeleton is on the outside of the bones'. It was difficult for Erin to disassociate the word 'skeleton' from the concept of bone. Through looking at books, reading and discussion, I felt her ideas were a little clearer. For her final entry in her 'think book', Erin wrote, 'what I now understand is that ants and insects don't have bones, but the skeleton is on the outside, and the outside is an exoskeleton'. This indicated that Erin was refining her understanding. The following statements summarise what some of the other children understood after their investigations.

> I understand that the crack was normal, because they have to have the crack there, because they need the space for their brain. — Fiona

> You can tell by the pelvic bone if it's a man or woman. — Jessica

> I understand that insects don't have bones. That people have 206 bones. — Denis

> I understand insects do not have bones, they have an exoskeleton. — Steven

> I understand that insects don't have bones. I think that shells are made out of that stuff that lobsters shells are made out of. — Vincent

Reflection

I could see that the children had acquired some knowledge about bones that they did not have before. They were keen to explore bones further, and their attitude to learning changed, even after such a short time. Denis, an unobtrusive member of the group made this comment: 'I like finding out about what *I* want to know, not what the teachers want us to know, that's usually boring'. This comment showed that Denis recognised this approach as something different and useful for him as an individual.

By working this way, I found that I was able to support the less able children. Steven, a child with learning disabilities, felt very much part of the group and was always encouraged by them. His feelings as a worthwhile member were demonstrated by his contribution to group discussions and group writing on the large sheets of paper.

I also noticed that the girls grew in confidence as time went by. I think that was because the atmosphere created by this approach made the group feel free and unintimidated.

Self-evaluation

I believe self-evaluation is an important part of examining the worth of the learning environment. I primarily achieved this by tape recording each session and analysing my performance. I asked myself the following questions.

- How well did I identify the children's ideas and questions?
- How effective was I in challenging, refining and extending the children's ideas?
- How well was I able to help the children resolve some of their questions about the topic?
- How effective was I in arousing the children's interests so that they continued to ask questions?
- In what ways did I help the children to develop their investigative skills so that they valued pursuing answers to questions?

I don't find it easy to teach science using the approach described in this unit. It is challenging because I need to research the area in depth and to be very well prepared. In addition, I need to think very carefully about how to interact with the children. I have to work on my questioning techniques – the tape recorder helps. This requires plenty of practice and time.

Overall, I believe that, whilst the approach is challenging, it does give me an indication of the level of interest children have in topics such as bones, how they are eager to ask and answer their own questions and how easy it is to explore a range of science concepts from such a humble start.

This chapter reveals the considerable learning that can take place when children's interests are followed. It is also an interesting chapter as an example of a teacher reflecting on her practice and finding ways to analyse her effectiveness. As Mary says, this is a very challenging way to teach science, but what the children gained in this unit was considerable. It is all too easy to get into a rut where we are comfortable with what we do. This work shows the value of reflecting on what we do, taking a few risks and trying something new. Face the challenge and try it yourself!

16 — What is the universe made of?
Jan Elliot

What is the universe made of? In this chapter, Jan describes how her eight- to ten-year-olds explored this interesting question. Jan first asked the children to brainstorm all of the things they understood about the make-up of the universe. This was followed by a carefully sequenced range of experiences that led to the children investigating the periodic table – what is it, what does it mean and how do scientists use it? The child-centred nature of the investigation is evident in Jan's explanation of how the children wanted to know more about the atomic weight of the items listed on the periodic table – they compared each item with gold, something about which they had some sense of weight.

Age	8–10
Situation	teacher with whole group, independent small groups and individuals
Science	using an interactive approach; taking into account children's existing ideas; bringing their ideas closer to the scientists' view, through investigation and exploration
Theme	introducing the children to the periodic table and its place in the scientific world

Introduction

Learners are always striving to make sense of their world and the world of 'chemistry' is an interesting area for children to explore. What understandings do eight- to ten-year-olds have about the make-up of the universe? How will they accommodate the idea of the universe being made of a little over one hundred elements? What sense will they make of atoms and molecules? To find out, a grade 3/4 composite class was observed as they undertook a topic of study, 'the make-up of the universe'.

Preparation

I purchased one large copy of the periodic table and smaller copies for each member of the class. Information about the table in books accessible to eight- to ten-year-olds was difficult. Encyclopaedias proved to be a good source of information.

Getting started

The first step was to discover what the children already knew about the make-up of the universe. They were asked to construct individual, simple concepts maps about the universe. As usual, when children are given the opportunity to demonstrate their prior knowledge about a topic, they showed it to be considerable. The expected things, such as earth, water, air and space were forthcoming, as the concept map by Brigit shows (Figure 16.1).

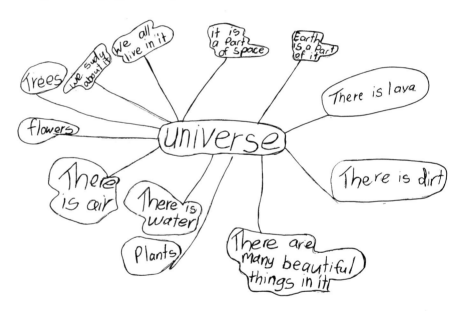

Figure 16.1 Concept map by Brigit.

Oxygen and carbon dioxide were not unexpected, and they appeared on several concept maps (Figure 16.2).

Some children, however, included hydrogen, helium and matter, and demonstrated a more sophisticated understanding of the make-up of the universe.

The class sharing session of the concept maps exposed individual class members to the ideas of the others. All the ideas were printed on cards for the children to illustrate. I colour-coded the cards according to whether they showed solids (black), liquids (green) and gases (blue), or a combination of these, and the children identified the nouns for all their ideas. They grouped the cards according to colour and were asked to theorise about my groupings. It did not take them long to come up with the answer.

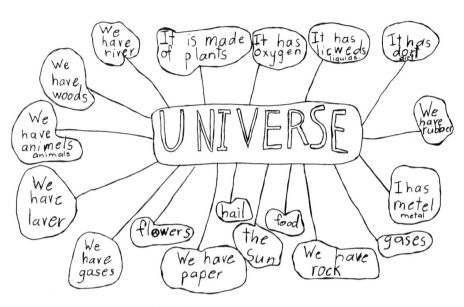

Figure 16.2   Concept map by Jennifer.

Explorations

We had thus established that the universe was made up of solids, liquids and gases. In order to understand that a mere hundred-odd elements make up the many things that the children have experienced, I felt that the children needed some conception of compounds. We began with mixtures of salt and water, sugar and water, oil and water, smarties and rice bubbles, and dried corn and sultanas. By shaking the oil and water, the children could observe the mixing and quite rapid separation back to the original components. Salt crystals soon began to form in the salty water, and the children predicted that the water was evaporating off, leaving the salt behind and that in time, the substances mixed would have separated.

Having established some understanding of mixtures, we moved onto compounds. First, we made 'slime'. Some polyvinyl alcohol had been dissolved in water before class (using 3.5 l of water and 90 g of PVA; boil the water, sprinkle in some PVC, stir and then simmer for half an hour, stirring all the time) as this takes some time. The children took 90 ml of the PVA/water solution and added food colouring. They added a pinch (0.5 g) of borax. Stirring began, and in about five minutes, we had a 'slimy' situation!

Figure 16.3 Mixing slime.

TAKE CARE

This slime exercise was followed by one to make 'stalagmites'. Groups of children dissolved 20 g of magnesium sulphate in a minimum amount of water and 20 g of sodium carbonate in a minimum of water. Then, pipettes were used to drip the individual liquids down a pipe cleaner, where they met and dropped to a piece of paper (Figure 16.4). The 'meeting' of the two liquids resulted in the formation of a compound, and stalagmites began to form. (It is worth pointing out to the children that natural stalagmites form in limestone caves over thousands of years – this classroom method is somewhat quicker!)

Figure 16.4 Making stalagmites.

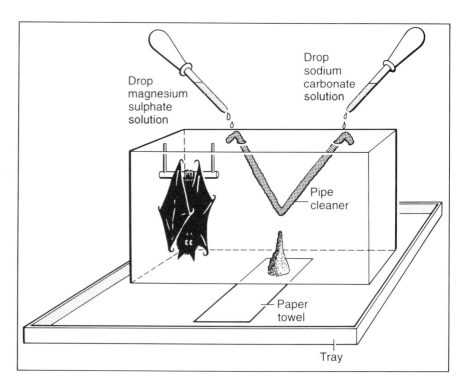

Considerable discussion followed these mixing activities, my aim being to develop an understanding that some things mix together and separate easily, whilst others mix together, but do not separate easily, or '... with our own power', as one student put it. We role-played being mixtures – smarties and rice bubbles – and we separated easily. I told the children about the polymer chains in 'slime', and we role-played being polymer chains. They 'experienced' the joining together and discussed how 'links' held the 'substances' together.

We discussed our class concept map in the light of our new understanding of mixtures and compounds. The children concluded that many things were compounds. Children introduced the terms and symbols H_2O and CO_2, picked up from the discussion. Seizing on this, I introduced the periodic table, on which we found hydrogen, carbon and oxygen. We discussed how the combination of these formed water and carbon dioxide. I explained that scientists had discovered that the elements displayed on the chart did, in fact, make up the 'matter' of the universe.

With no further discussion, each child was given a copy of the periodic table (Figure 16.5) and there was a lot of noise as children read and discussed these. A larger copy of the table was displayed on the wall of the classroom.

I	II											III	IV	V	VI	VII	0
					H (Hydrogen, 1)												He (Helium, 2)
Li (Lithium, 3)	Be (Beryllium, 4)											B (Boron, 5)	C (Carbon, 6)	N (Nitrogen, 7)	O (Oxygen, 8)	F (Fluorine, 9)	Ne (Neon, 10)
Na (Sodium, 11)	Mg (Magnesium, 12)											Al (Aluminium, 13)	Si (Silicon, 14)	P (Phosphorus, 15)	S (Sulphur, 16)	Cl (Chlorine, 17)	Ar (Argon, 18)
K (Potassium, 19)	Ca (Calcium, 20)	Sc (Scandium, 21)	Ti (Titanium, 22)	V (Vanadium, 23)	Cr (Chromium, 24)	Mn (Manganese, 25)	Fe (Iron, 26)	Co (Cobalt, 27)	Ni (Nickel, 28)	Cu (Copper, 29)	Zn (Zinc, 30)	Ga (Gallium, 31)	Ge (Germanium, 32)	As (Arsenic, 33)	Se (Selenium, 34)	Br (Bromine, 35)	Kr (Krypton, 36)
Rb (Rubidium, 37)	Sr (Strontium, 38)	Y (Yttrium, 39)	Zr (Zirconium, 40)	Nb (Niobium, 41)	Mo (Molybdenum, 42)	Tc (Technetium, 43)	Ru (Ruthenium, 44)	Rh (Rhodium, 45)	Pd (Palladium, 46)	Ag (Silver, 47)	Cd (Cadmium, 48)	In (Indium, 49)	Sn (Tin, 50)	Sb (Antimony, 51)	Te (Tellurium, 52)	I (Iodine, 53)	Xe (Xenon, 54)
Cs (Caesium, 55)	Ba (Barium, 56)	La (Lanthanum, 57)	Hf (Hafnium, 72)	Ta (Tantalum, 73)	W (Tungsten, 74)	Re (Rhenium, 75)	Os (Osmium, 76)	Ir (Iridium, 77)	Pt (Platinum, 78)	Au (Gold, 79)	Hg (Mercury, 80)	Tl (Thallium, 81)	Pb (Lead, 82)	Bi (Bismuth, 83)	Po (Polonium, 84)	At (Astatine, 85)	Rn (Radon, 86)
Fr (Francium, 87)	Ra (Radium, 88)	Ac (Actium, 89)															

Figure 16.5 Periodic table.

Class sharing at the end of this session produced a list of elements with which the children were already familiar – hydrogen, calcium, chromium, iron, nickel, copper, zinc, oxygen, helium, carbon, neon, aluminium, arsenic, krypton (Superman!), iodine, tin, lead, radon, silver, platinum, mercury, radium and uranium. We decided to make a collection of elements we could find in their natural states.

We collected aluminium (foil and a block from melted cans), lead (fishing sinker), copper (wire), zinc (from dry-cell battery covering), chromium (from a car mirror) and magnesium (from inside a flash cube).

Encouraging the children to ask questions

I told the children that they were going to make individual books about elements. Class discussion about what they wanted to find out about elements led to a rough format for the books. The children suggested:

- Who found the element, how and when?
- Solid, liquid or gas at room temperature?
- Its weight.

This caused a lot of discussion. The *Time-Life* large periodic table included 'atomic weight' in the information about each element, and the children noted that hydrogen, the first element on the table, was the lightest and that the weight of the elements increased as the table went on. Some decided that finding gold on the table and finding 'their' element in relation to this on the table would give them some idea of its weight – heavier or lighter than gold. They then came up with a few more questions:

- What is its symbol?
- How did it get its name?
- In what things is it present naturally?
- In what people-made things is it present? Why?
- What colour is it?

The children chose an element to study, and we listed these to ensure that no element was chosen twice.

Specific investigation

A book format, including all the items listed above, was photocopied to aid the collection of information. The children were involved in a language topic at the time, so they decided on which characteristics their books should have: table of contents, facts, non-fiction, true illustrations, diagrams or photographs, sub-headings and an index.

The children discussed how they were going to get the information they wanted. They decided upon using the periodic table, 'experts' and library books, including encyclopaedias. Research began, and the end of each session concluded with the children sharing what they had discovered about 'their' element.

This kind of 'review' discussion helps to keep children aware of what others are doing and helps the general enthusiasm and flow of talking and finding out.

Publishing their discoveries

When most of the children had completed their research, publishing decisions had to be made. What would we call the set of books? What publishing symbol would we use? What blurb would we put on the back cover? Over a few days, the following names were presented for consideration for the title of the series:

Exciting Elements
Finding Out Books
Fun in Science Books
Did you know that . . . ?
Science Corner
Wonders of the Universe
Elementary Things
Question and Answer Books.

The title 'Wonders of the Universe' won the day. (I personally thought the name 'Elementary Things' was perfect!) For the blurb, we decided on: 'Look inside and you will see a world of science'.

Children's reflections on their learning

At the conclusion of the unit, the children were again asked to produce a concept map of what the universe is made up of. Leila's effort is shown in Figure 16.6.

Figure 16.6 Leila's concept map.

Some of the original ideas remained, but the following shows considerable development in the children's ideas.

> The universe is made up of light and sound, matter, atoms, solids, liquids, gases, protons, electrons, 105 elements, neutrons, hydrogen, oxygen, lithium beryllium, sodium, magnesium, potassium, calcium, titanium, iron, nickel, copper, zinc, argon, helium, krypton, neon, gold, silver, carbon, chlorine, lead, sulphur, silicon, platinum, fluorine, thallium, aluminium, oxygen, mercury, copper, tin. The elements have different colours, they are all in something that people have made, all elements have weight, all elements are made out of atoms, they all have a state at room temperature, they are found in their natural state and all elements have a symbol.

Transferring knowledge to other topics

Incidental comments and classroom discussion following the unit on the periodic table revealed the acceptance of the concept of elements making up matter. When the children poured vinegar into baking powder dissolved in water with little pieces of spaghetti dropped in, bubbles of carbon dioxide formed on the spaghetti pieces, causing them to rise to the surface, lose their bubbles and sink again. The children concluded that the bubbles were caused by a 'combination' of substances and said they believed it was carbon dioxide because baking powder is sodium bicarbonate. They also said that hydrogen and oxygen were there – in the water.

Conversations about solids, liquids and gases and their properties have been noticed in other topics. When the class topic was 'air', two children revealed considerable knowledge, a development from their study of elements. Figure 16.7 shows a page from Leigh and Daniel's 'thinking book', in which there is an understanding of air consisting of nitrogen, oxygen, argon, neon, helium, krypton, hydrogen and xenon, and the percentages of these present. The page shows that the children have understood that air is a mixture.

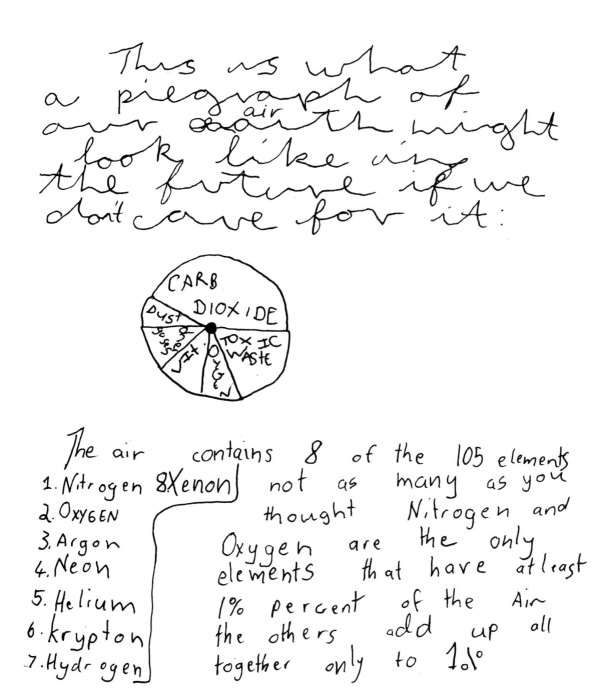

Figure 16.7 Extract from Leigh and Daniel's thinking book.

You can see from this extract how valuable a 'thinking book' can be in prompting children to explore their own ideas and to extend these ideas when interacting with others.

In the concept map in Figure 16.8, Katie shows her understanding of the elements in ozone, that oxygen is a molecule containing two atoms and that a chemical process is happening in the formation and destruction of ozone.

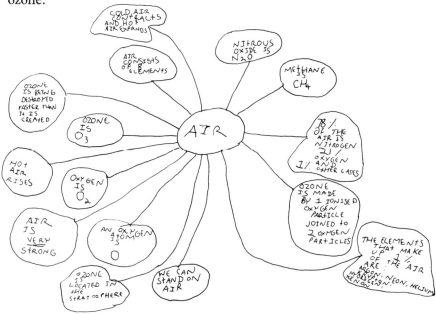

Figure 16.8 Katie's concept map of 'air'.

Conclusion

The idea of the elements making up the universe did not appear to cause conceptual difficulty for the children. They worked enthusiastically throughout the unit and displayed considerable knowledge about elements at the conclusion. Children aged between eight and ten already know a great deal about how their universe works. From a study of the elements on the periodic table, it is clear that they are able to accommodate the idea of different combinations of basic elements being responsible for the 'matter' of their universe.

An adventurous choice of topic! But clearly one that does manage to interest the children and develop their knowledge. It also manages to make links to other curriculum areas. For instance, it is easily integrated into our class 'publishing' and in that sense, we can think of our science as part of a 'whole language' curriculum – as science with reason.

17 — Rock week
Jill Jesson

This interesting chapter shows some children exploring geology, starting from their own questions. It shows how children devise their own investigations and record for themselves in a variety of ways. This work took place over just one week, and the work that the children do is impressive, showing the teacher's ability to encourage her children to set about answering their own questions.

Age	10–11
Situation	teaching a mixed-ability class of children, using their queries as a starting point; the children worked in groups of four or five to answer further questions which would lead them to solve their original problems
Science	geology: investigating rocks
Theme	Children controlling their own learning and taking responsibility for directing the methods of investigation and the recording of their results.

Introduction

I often start work with my class by asking what they would like to discover. But is not just a question of finding out what they *think* they know, but ascertaining what they have actually assimilated, including what misconceptions they have developed. As Asubel said in 1968:

> 'The most important single factor influencing learning is what the learner already knows; ascertain this and teach him accordingly.'

I knew that my class had done a term's work two years ago on the theme of 'A journey to the centre of the earth', and this had included some work on geology. While we were studying 'Materials', it seemed appropriate to look at this area again from a new angle.

When teaching older children in the junior school I like them to direct their own education whenever possible. I believe that children should learn to ask their own questions and then go about supplying their own answers, since this increases motivation and gives them some responsibility for running their daily lives. I see my job as an 'enabler' who will allow them to discover what they can, as well as an 'educator' who will lead them to new questions, solutions or answers. This does not mean that they are just left to discover everything for themselves, since this is not possible or practical. But I do feel that giving them greater responsibility and control over aspects of their education is a good way of leading them into adolescent and adult life, with its associated responsibilities.

What do you know?

Questions about rocks and minerals had already arisen during the term while we were looking at the properties of a variety of materials. The children had grouped materials into natural and man-made items, those of animal, vegetable or mineral origin, and those which were alive, once alive or never alive. Rocks kept cropping up.

I explained to the children that I was planning to spend one week looking at rocks and that I needed to know what they already knew about rocks and what they would *like* to know about them. They were asked to draw concept maps or to write me a note so that I would be able to see where to lead them. At the end of the week, they wrote a second set of maps or notes to show what they had learned by themselves and from their friends. Figures 17.1(a) and (b) show one girl's concept maps at the beginning and the end of the week.

Discussion during this writing brought out a number of points. First, although they had done quite a lot of work on this subject previously, some of it had been remembered incorrectly or misunderstood in the first place. Parts of one concept seemed to have been tacked on to another. Several children remembered the words 'igneous', 'sedimentary' and 'metamorphic', but only one boy had any idea what they meant. Furthermore, they did not know the difference between a rock, a mineral and what they described vaguely as a stone. Various other interesting misconceptions were unearthed:

> 'The hardest rock is marble
> and the softest is clay.'   Liz

> 'Some rocks can be on fire,
> like in a volcano.'   Tom

> 'Methamorthic is sedentery or
> igniose that has been changed
> by the weather.'   Bobbie

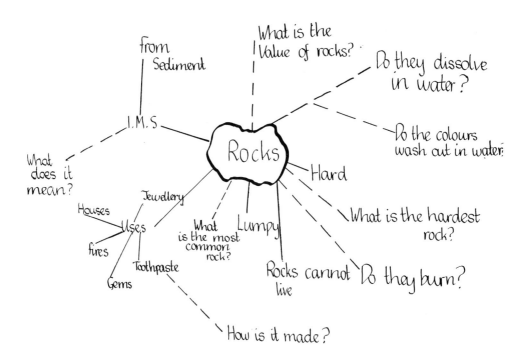

Figure 17.1a Emily's concept map 'before'.

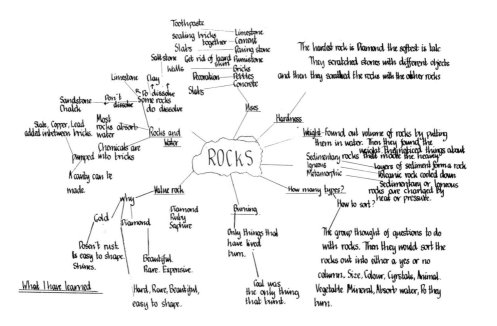

Figure 17.1b Emily's concept map 'after'.

> 'All rocks are made by crushed-together trees and plants. After hundreds of years, the plants and trees will form rocks.'
>
> 'Some rocks come from under the sea, but most rocks come from under the ground.'   Laura

Alison's note was typical of the rest:

> Rocks can be all different colours, shapes and sizes. There is different kinds of rock like sedimentary rock, Blue John and limestone. Rock is formed by leaves, coal, mud and things like that all crushed together. It has to be left for a long time for the rock to produce. Some rocks are rough and some are smooth. Rocks can be all sorts of shapes. Some are jagged and some are straight. They can have different values. Some are very expensive. Rocks are used for toothpaste, houses and things like that.

Alison had remembered quite a lot of facts, but some of them had got rather confused with each other. Like most of the class, she was not sure of the difference between a rock and a mineral, or the millions of years involved, but she has a good idea of their tremendous variety.

What would you like to know?

There was much enthusiasm for investigating rocks, and several children had already brought their collections of rocks and minerals in to make a display for the rest of the class. They had no difficulty in deciding what they now wanted to know, and the following represent only a sample of the questions they came up with.

> 'I want to know what is the most valuable rock, gem or mineral and why it is.'

'I would like to know what 'metamorphic' and 'igneous' mean and what they are formed by.'

'Do some rocks and minerals burn?'

'I wonder how long ago rock was made ... did anyone make rock?'

'Does water make rock crumble?'

'I would like to know what rocks are used for.'

Working with their ideas, I listed the eight areas which most children requested information on. These were:

- How hard are the hardest rocks?
- Do rocks burn?
- Which rocks are the most valuable?
- Which is the heaviest/lightest rock?
- Do rocks soak up water or dissolve away in it?
- What is 'igneous', 'sedimentary' and 'metamorphic' rock?
- How many types of rock are there?
- What are rocks used for?

Other questions would be dealt with as they arose. It would not be possible in the time available to have every child answer every question, so I decided to give groups of three or four children one question to research, on the understanding that they would share their answers with the others at the end of the week. I find that, if children know they have to teach others what they have learned, it helps them to clarify their ideas and gives an added incentive for them to learn. It is also a good way to assess what has been learned if they can teach peers concepts they have newly acquired. Before they began, we clarified the differences between rocks and minerals and discussed the meanings of words like 'stone', 'pebble', 'jewel' and 'gem'. I encouraged them to use dictionaries for an initial definition of these words and then to seek clarification in encyclopaedias and reference books. Some discussion was still necessary for those who were unsure of various points.

On the Monday, each group was given an envelope with their question to be answered on the outside. Inside was the first of three more questions which would direct them towards an answer. They received a new, related question on the Tuesday and another on the Wednesday; they then spent some part of Thursday recording their results in whatever way the group thought appropriate. Friday's rock time was given to a short presentation of discoveries by each group. Although some problems required more practical work than others, I attempted to give each group:

- one question which would require hypothesising and practical testing of rocks;
- one which would require some research from the school library;
- one which needed creative thought and application of what they knew to find the answer.

The envelope idea was an approach which I had not tried with this class before, and they were eager to open them. It was necessary to balance the practical and research questions so that not everyone was using the same rocks at once, but I had enough for four groups to work at a time with at least eight samples of rock each.

Answering questions

How hard are the hardest rocks?

- Can you list these eight rocks in order of hardness?
- What are the hardest and softest rocks in the world?
- What uses for rock can you find which are connected with their hardness?

Early in the discussions, John hit on the idea of scratching each of their samples with all the others to see which order they could put them in; but since he had a low status within the group, his idea was rejected until they had tried various other methods. Dropping them onto paper to see the damage was ruled out as too difficult to make into a fair test, while scratching them with other items, such as a nail or coin, only sorted a few from the rest. Only when these ideas had failed and they had discussed the differences between weight and hardness did they accept and quickly solve their problem with John's original idea. They had some problems understanding the concept of Moh's scale, since they felt it must be the 'top ten' hardest rocks, or at least the top five and bottom five. It took a while for them to understand that it represented a sliding scale of hardness. They found lots of uses for rock which surprised them and were delighted to discover it in toothpaste and a pumice stone (Figure 17.2).

Do rocks burn?

- Which rocks burn? What do you notice when you try with these samples?
- What sorts of materials do burn?

Figure 17.2 James's list of uses for rock.

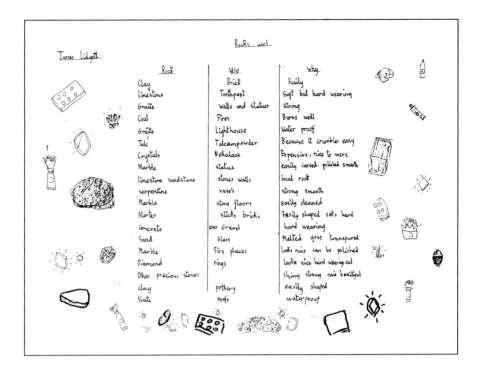

- Are rocks on fire in a volcano? What can you find out about this?

The group enjoyed trying to make a variety of rocks catch fire. They didn't expect many of them to ignite, but some thought that coal, lava, slate or chalk might. Only the coal caught in the two minutes they had allowed, and so they turned their attention to materials that did burn. They listed some and tested others and, with careful questioning, came

Figure 17.3 Mark's tests to show which materials burn.

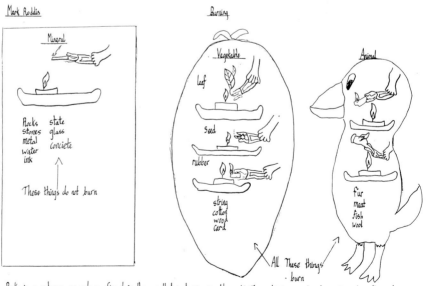

141

to see that only things which were alive, or had once been alive, would burn. This led on to a discussion of the carbon cycle. By this time, they were sure that rocks in a volcano were not really on fire but only melted, and they looked in books to discover that it was the gases in the rocks which were burning, rather than the minerals themselves (Figure 17.3)

Which rocks are the most valuable?

- What makes a rock or stone precious?
- Colour is one way we can tell one rock from another. Scientists sometimes use a streak test. Can you do some? Which was your most surprising result?
- Crystals come in many regular shapes. What could you do to show this to the rest of the class?

This group was the most difficult to set tasks for, since we had no gems at hand for them to do practical tests on. They began with a research-and-thinking problem which required them to identify just what made a precious stone precious. The question of colour had arisen here, so I linked this to a question posed earlier on whether minerals were always the colours they appeared to be. They were intrigued with the streak test idea and tested many samples to see whether the theory held true. Finally, they enjoyed attempting to model some of the crystal shapes they had discovered in card or clay (Figure 17.4).

Figure 17.4 Clare's concept map of precious stones.

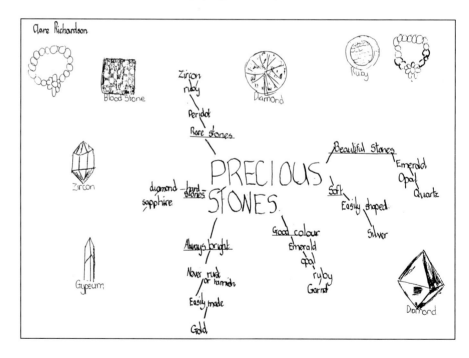

Which is the heaviest/lightest rock?

- Can you arrange these rocks in order of weight?
- How could you compare their weights with their volumes?
- What do the heavy ones look like? Can you see why they are so heavy?

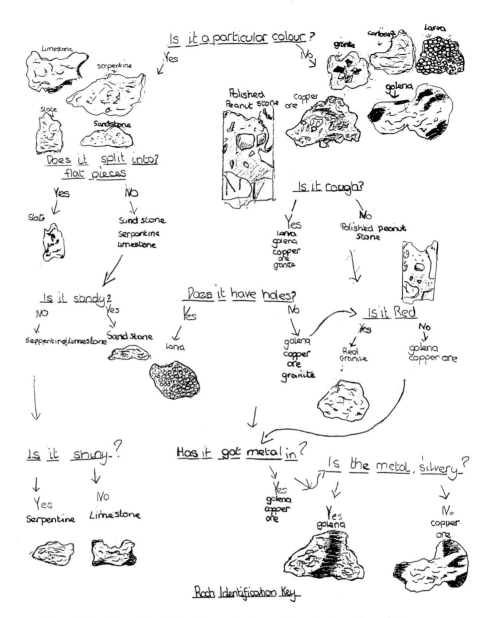

Figure 17.5 Mary (age 11) used actual rock samples and a large sheet of paper to move the rocks around and sort them.

This group was composed of three bright children who swiftly arranged their samples in order of weight and, when questioned, immediately pointed out that it was an unfair test due to the differences in rock size. They hit on the idea of water displacement to measure the volumes and calculated the relationship between the two figures for themselves. Their problem was one of accuracy when measuring, and it required several tests and retests to collect accurate results. They came up with many theories as to why one sample was heavier than another, including the 'compactness of the particles of it', the mineral content, whether it had holes in and 'whether the bits of mineral which make it are light or heavy' (Figure 17.5).

Do rocks soak up water or dissolve away in it?

- Do these rocks absorb water?
- Do any rocks dissolve in water?
- If you were building a house, how could you stop rainwater seeping into it?

This group were sure that most rocks soaked up water. When they reweighed their samples after a four-hour soak, they were surprised only by how much more some had absorbed than others. Just slate and flint were left untouched, which led naturally on to researching how to keep houses free from damp. They all knew that salt dissolved in water, although not all recognised that it was a rock, but were confident that clay, chalk and sand would do the same. Work they had already done on saturated solutions and suspensions helped to disprove some of their theories, and they were left with limestone, which one girl knew was dissolved by streams in the Peak District to form caves.

What is igneous, sedimentary and metamorphic rock?

- What do these words mean?
- Can you devise a test to show how rainwater affects the landscape?
- Compare granite, basalt and lava, which all come from volcanic rock. What do you notice?

This group looked at the long words they remembered from two years ago. They began by researching their answers, and then built a hill of stone, sand and soil which they sprayed with water to study the erosion and deposition of materials. The resulting delta in the lake below their mountain clearly showed graded bedding, and the word 'sedimentary' had real meaning for the first time. They looked at several types of igneous rock under the microscope and were confused as well as intrigued to find such variety. We discussed how other materials altered when they were scorched, melted or squashed, which gave them some idea of how metamorphic rocks were just altered variations of the other two types.

How many types of rock are there?

- Sort the rocks you have according to their characteristics. Can you devise a branching key to sort them with?
- How many other ways could you sort rocks?
- Find a table for classifying rocks in a book and try it on your samples.

These children began by playing 'Guess which one I am thinking of' to practise sorting their rocks by visible characteristics. They then composed a branching key to sort them and compared it with a key in a published book. They found twenty ways of separating them (Figure 17.6).

What are the rocks used for?

- How many uses for rock can you find around the school?
- Look at the properties of each one and say where it comes from and why it is used for that purpose.

Figure 17.6 Jessica (age 10) found twenty ways to sort her rocks.

- Can you sort these building materials into those which are most hard-wearing and best for building?

This group began by listing all the uses of rock they could see around school. They looked at the properties of each, whether it was natural or man-made and said why they thought it was used for that purpose. Having compared the pros and cons of various other materials for building, their research was recorded in table form. Two of them found it difficult to see clay as a soft rock, and thought it was just a hard soil, since one had a father who complained about the clay soil in his garden. They therefore said that bricks were soft because they were made from soil and were surprised how hard some were when compared with local sandstone.

Cross-curricular activities

While this work was in progress, normal lessons in maths, English, PE and other curriculum areas continued as well. In art, they drew some of the samples and tried to represent them in collage and clay. One boy who was intrigued by the enormous time scales involved wanted to make a chart to show this. I suggested using string, which he measured out using 1 mm to represent 1 million years. The resulting length of string was 4.6 m, and he coloured it with inks to show when the first plants, animals, dinosaurs and humans appeared. When this was held up to show the rest of the class, all were impressed by the length of time when 'nothing much was happening but weather' and how humans appeared only in the last 2 mm. Later, he coiled it up onto card and labelled the coloured sections. Other children chose to paint and write about some of the samples.

Conclusion

I often get my classes to tell me how they feel about the way I have presented a subject to them so that I can adapt and vary my approach to suit their ideas and aptitudes. They like to know that their tastes and feelings are being considered and that they are being treated as individuals, not just an anonymous cohort of kids. The verdict on the envelopes idea was one of unanimous approval because, as Lindsey said, 'It was exciting not knowing what the next question was going to be'.

On the Friday, each group reported to the others about what they had learned and answered questions from their peers. Although, as always, a few were tongue-tied, at first most coped quite well and enjoyed putting on a bit of a performance. In any case, it was a good way to share experiences and for me and them to assess what they had learned, for, as one child remarked afterwards, 'Now that I've said it, I know what it is I mean'.

Jill's way of working shows how powerful it can be when children are directing some of their own learning. The different approaches of the groups reveal a considerable level of independent thinking.

What comes across strongly in this and all the stories in this section is how children can be fascinated by rocks, space, their bones, their environment, etc. Each of these stories shows how that interest has been steered into the same good investigative work based on the children's own questions. In many cases, there is clear evidence of substantial learning. Science taught in this way is clear and exciting – both for the pupils, and as this writing shows, for the teachers.

Section D A practical approach

For all of us, teaching science requires some kind of change and development

—18— How can I organise myself?
Sue Atkinson

This chapter pulls together some of the themes of classroom organisation that have emerged in earlier sections.

Classroom organisation

One way of organising groups for practical science work is discussed in Chapter 4, in which the 'plan, do and review' structure is used. Many of the other chapters build on scientific ideas raised during discussion times with children, and both of these ways of organising point to the crucial role of language and communication; however, my problem is always how to keep track of all that in a busy classroom.

The science investigation board

In Chapter 4, I described how I used a display board to pull together the work and to provide a focus for questions and ideas. (I also did this on the maths board and I did a similar thing on the book review board in the reading corner.) I put up questions, ideas and starting points. Over the years, I have worked with colleagues to improve these boards; Figure 18.1 shows one that several of us like.

As a guide to what we did, this type of board worked reasonably well, but I was left wondering if it might limit thinking rather than expand it. I could refer to it as I worked with groups, and I thought that the nine-year-olds did refer to it, but for the younger ones, the excitement and meaning was in the practical work and in displaying their work. I rarely observed them using the board as guidance, though I did refer them to it when I thought it would clarify what they were doing.

Sometimes I find it very helpful to have somewhere in the classroom where children can put up their questions. This can be on a science board, or in a class book, or simply jotted down on a temporary easel put up specifically for a few days to get the questions flowing. Whatever method we use of accessing the children's questions, the really important thing is that children need to feel that what they want to explore is important to us to.

Figure 18.1 Science board.

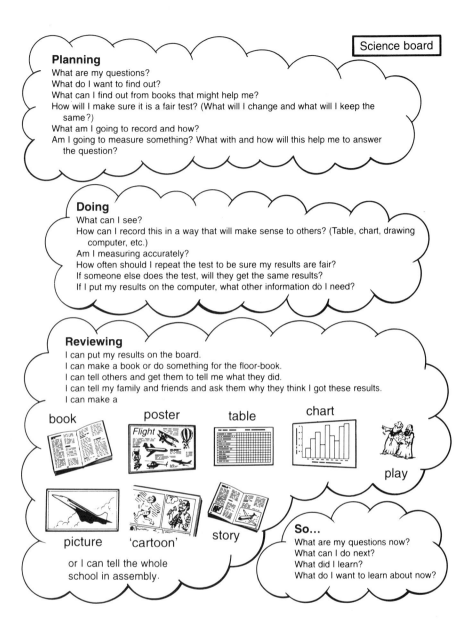

Children recording their work

Throughout the book, we have given examples of children's own recording of their work. We found that it was through these recordings and by listening to and observing children that we could try to access what they were thinking and what sense they were making of what we were doing. The use of some kind of concept map – developed by individuals or by the class group – is crucial in finding out at the start of a project what the children already know and believe.

Some children can develop extraordinary theories about their work. A colleague told me of a child who was wrapping onion skins around eggs to dye them brown. He had obviously been told he was 'dyeing' eggs, and when asked by my friend what he was doing, he replied 'killing eggs'! I think that it is only as we listen, observe and study children's recordings that we access this type of thinking.

It can be helpful to use a chart designed to help children to be systematic in their recording of their science. Many teachers I work with use charts like these, and for the young child, they can be a way of recording their work *as they go along*. (I think we can switch children off 'work' if we say to them as they complete a task, 'now go and write about it'!)

Name _____	's results
I am trying to find out . . .	I tried to find out . . .
I think this will happen . . .	I changed . . . / I kept . . . . . . the same
. . . because . . .	I measured
I found out . . .	I am going to show others my results in . . .

Figures 18.2a, b, c, d   Four examples of recording sheets.

Investigation plan by_____	I need . . .
My question is . . .	
I will measure . . .	I will record my results . . .
I will change . . .	
I will keep . . . . . . the same.	I think this will happen . . .

Figure 18.d

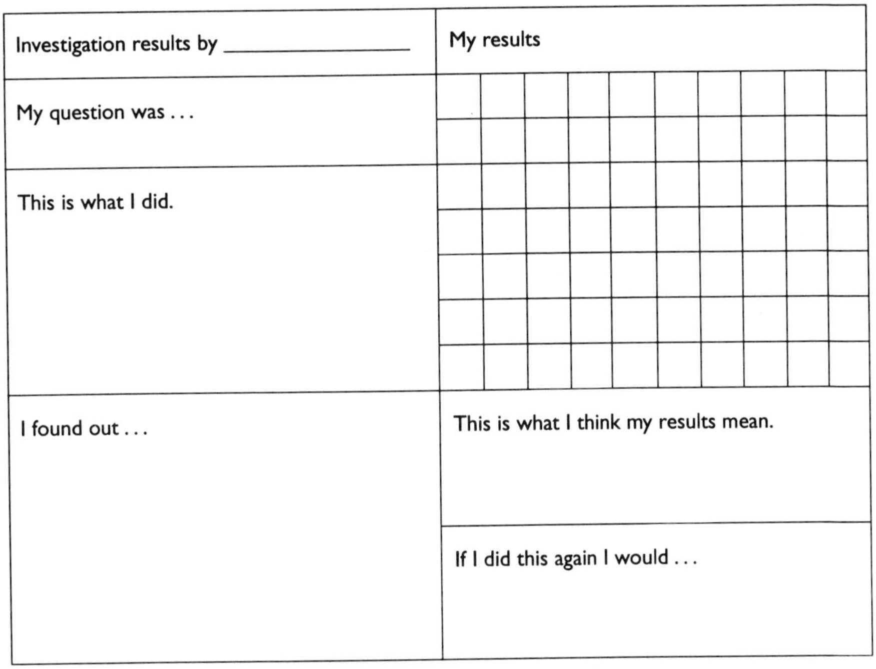

Investigation results by _____	My results
My question was ...	
This is what I did.	
I found out ...	This is what I think my results mean.
	If I did this again I would ...

I would not expect a child under eight to use a chart like the ones shown in Figures 18.2a, b, c and d each time they do science investigations. There are many other ways to record, and we may well find that we can access the child's thinking most effectively through their own unstructured recordings. However, there would seem to be a particular value in using charts when children are working on an investigation they have planned themselves. There are a number of charts like these, and you might want to copy them to use in your class. Figures 18.2(a) and (b) might be better with younger children and (c) and (d) with those over seven or eight. As you use them, you will want to adapt them and encourage your children to plan their own recording sheets.

Other ways of recording mentioned in previous chapters are:

- floor books or big class books;
- concept keyboard (especially useful with under-sevens);
- zig-zag books or some kind of individual booklet;
- a diary (particularly valuable for recording changes, e.g. the hatching chicks or growing beans);
- 'cartoons' (paper folded into four, six or eight sections);
- pictures (annotated pictures are particularly powerful, as they demonstrate some of the children's thinking);
- posters;
- making some kind of dramatic production or writing a story;
- planning an assembly;

- teacher/child and child/child discussions in the classroom.

Even very young children can record with, perhaps, an annotated picture using 'developmental' writing. Recording helps us to assess understandings and to see these developing and gives us a starting point for discussion; it also helps the children to express their ideas. That is often the starting point for clarifying thinking.

The classroom layout

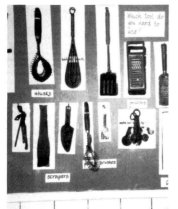

Figure 18.3 'Tool shadows' ensure equipment is returned to the correct place.

Planning for science, maths and technology requires space and thought about the most efficient use of that space. For many of us, there are considerable problems. Our room seems small, badly designed, or, as I had once, is entirely carpeted, with glue, water and anything messy being banned by the caretaker!

What is crucial for the children is to have free access to a variety of equipment that will enable them to explore what they are interested in. Much of this equipment is 'junk', and we find it best to store this in clearly labelled boxes.

Maulfry Hayton has clear 'shadows' of tools to ensure that they are replaced after use (Figure 18.3).

In an early-years classroom, a layout like the one of Maulfry's classroom (Figure 18.4) gives space and value to science as a crucial area of early learning. The layout may be different in a middle-years room, but there is still the need to give science the status it needs as we approach the twenty-first century.

Figure 18.4 A classroom set up to give space for children's investigations.

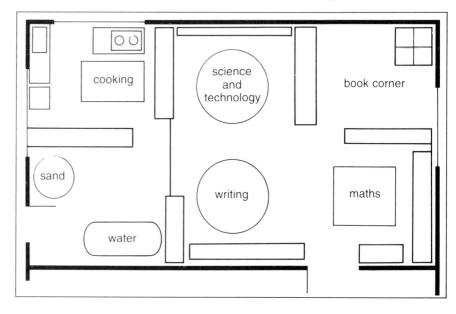

152

Making a science garden

As we plan for science, much of that learning will need to take place outside the classroom, and some schools have used their grounds to make some provision for science. The space doesn't have to be large. Slopes for running cars down can be constructed against the school wall, and tubs of plants in various soil types can be a focus for a bleak playground (Figure 18.5).

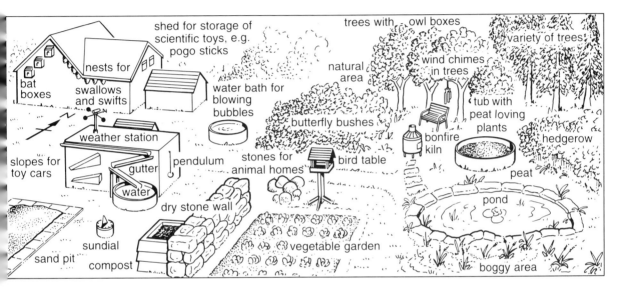

Figure 18.5 Science in the school grounds.

What is my role?

As is clear from the stories in this book, the teacher's role in the teaching of science is a complex one.

1 We need to find out what children already know and how they understand what they are doing. (What did the little boy make of his teacher telling him to 'kill' eggs?)

2 We need to find ways to help children to communicate their ideas so that we can 'get their thinking out on the table'. If a child thinks the magnet works 'like glue – it's sticky' (Vijay, age 6), then we need to know that and help children to develop more 'standard' scientific ideas.

3 We need to plan our classroom so that there is plenty of opportunity for children to explore and try out their ideas. This needs time, discussion and access to clear books that will inform the children. There are many issues here about how we organise ourselves; we must plan for equal opportunities and be skilful in helping children to raise questions (see how Rosemary Stickland does this in Chapter 6).

4 We need to listen to our children and encourage them to keep challenging and questioning their results and ideas. It is this process of refining ideas that helps children to change their thinking towards something more 'scientific', and as we listen and observe, we can assess this process. (See Chapter 20 for assessment.)

5 We need to keep working at our interventions! I tape myself at least once every half-term as I work with a small group. This is always desperately embarrassing, but no one else hears it, and I really do think it has helped me to become much more challenging and careful about what I say. Somehow we have to strive to be open-ended enough not to be thought of by the children as always having the 'right' answer. I find it is not just what I say, but the tone of my voice that is important. (For some examples of open questions, see Chapter 12.)

6 We need to develop ways of getting children to question and explore. Rosemary Stickland shows one way to do this in Chapter 6, and in Chapter 12 I show how I use brainstorming to get ideas going. Whichever method we use, the aim is to get at what the children already know and what they want to find out about.

7 We need to use our own scientific knowledge to assess the progress the children are making and direct them towards fruitful lines of enquiry and challenge their thinking. Sometimes we will have to tell children this information, e.g. names of animals, rocks and what makes a test 'fair'!

How can I develop my confidence?

Many of us feel some lack of confidence in science. We need ideas for ways to develop science with children and some guidance on scientific content. We also need to feel enough confidence to 'let the children go' and explore their own ideas. In my experience and when I work with teachers, it pays off to learn to say to the children, 'I don't know – let's find out'. We really don't need to know the answer to every question that children will raise. We need to read and find out about the 'content' – for example, the facts about forces (and we have put some suggestions to help with this in the resources list at the back) – but it is our way of questioning and our classroom practice that I think is more important than knowing all the facts. It's not so much what we know that will help us to be good science teachers, but observing and listening to our children and the ways in which we plan for them to explore and ask questions for themselves.

You could use this chapter as a starting point for discussing science teaching in your school with parents and colleagues. Does it make sense to approach the teaching of science this way?

19 — Writing a primary science policy
Graham Peacock

In this chapter, Graham guides teachers through the complex process of writing a science policy document. He shows how it can be both comprehensive and simple – a 'user-friendly' document.

Introduction

Schools need to produce policy documents and schemes of work for every area of the curriculum. These should be useful working documents which make the job of teaching easier and more rewarding. But all too often their production becomes a burden, and the resulting paper does little to help teachers.

Before starting work on either a policy or a scheme, you need to be clear about the differences between the two documents. The policy is the general framework and overall approach of the school to the curriculum area. The scheme of work gives details about what science is taught and seeks to ensure progression and avoid repetition.

Moving towards a written science policy

There is no single recipe for making a science policy, but you may find the following basic principles helpful:

- A policy should be a short document. If it is longer than two sides of typed A4, people are unlikely to read it.

- The document needs to be updated regularly, and all the teachers on the staff should play a part.

- Avoid jargon and use everyday language. Don't repeat what is in the National Curriculum documents, but make brief references to it.

Approaches to planning

The way in which you go about planning your science policy will greatly influence the eventual efficacy of the policy. The whole staff must be involved to some degree in writing the document so they feel that they 'own' it. However, not all the staff will need to be involved at every stage. Starting with the least involvement and working towards maximum involvement, the options are:

- The policy can be written by the head teacher and the science co-ordinator. The staff could comment and their comments be incorporated into a revised document. This is quick and efficient, but it is unlikely that staff will feel that it is their policy.

- A skeleton exemplar science policy could be written by the head and the co-ordinator. The staff could then use this as a basis for their own comments and amendments. This has the advantage of speed and simplicity. If done with care, staff can feel that the resulting document is theirs.

- You could present a list of headings which you feel are important to include in a science document. The staff could then work in groups to write the document collaboratively.

- The staff could be asked to list the points they want their science policy to cover. They could then work collaboratively as above. The disadvantage of the last two methods is the time and energy required.

Possible headings

1 Why we teach science at our school.
2 Aims of our science teaching.
3 The National Curriculum Programmes of Study or other curriculums, Statements and Profiles.
4 How we plan our science.
5 Continuity and progression.
6 Teaching methods.
7 Equal opportunities and differentiation.
8 Resources for science.
9 Safety in science activities.
10 Assessment policy.
11 Record keeping.
12 The role of the head teacher.
13 The role of the science co-ordinator.
14 Local Education Authority documentation.

Evaluating the policy

Once the policy is written, read it with the eye of someone joining your staff for the first time. Does it give the flavour of science at your school? Is it accessible for governors, (or school council or school board) or interested parents?

A specimen policy follows for you to criticise and improve on. You may decide to start your staff discussions with this structure in mind. Your own policy should honestly reflect what you and your staff see as important. Try to avoid simply repeating the current orthodoxy which may end up being meaningless and please no one in the long run.

A SAMPLE PRIMARY SCHOOL

Draft Science Policy

1   Why we teach science at our school.
Science makes an increasing contribution to all aspects of life. Children are naturally fascinated by everything in the world around them, and science makes a valuable contribution to their understanding.

2   Aims of our science teaching.
We aim to:

- teach the children scientific skills;
- teach the children scientific knowledge;
- build on the children's natural curiosity;
- stimulate them to investigate and question.

3   The National Curriculum Programmes of Study Statements and Profiles.
We base our teaching on the programmes of study for key stages 1 and 2 (level 1 or 2). The attainment targets inform our planning and particularly help with ensuring progress. In addition, we use the level descriptions to monitor the progress of individual pupils.

4   How we plan our science.
We plan from topics. This ensures that our science contains cross-curricular links and that science is not done in isolation. Science has important links with technology, maths and geography, but we do not neglect the excellent opportunities afforded for language development in science. Where science mini-topics are done, any links with other curriculum areas are exploited. We try to plan our topics as year groups and build on materials kept in a central file.

5 Continuity and progression.

We plan our topic coverage as a whole school. Year groups then use topics as a basis for planning schemes of work. Our plans are filed centrally so that others may use them to see what experiences the children have had. We keep and pass on whole-class records and also records for individual children. The top year in each key stage liaises with the relevant teachers in the next school.

6 Teaching methods.

We believe that the best way to learn science is through first-hand experience. However, children should be helped to make sense of this experience through discussion and application of their learning to new situations.

7 Equal opportunities and differentiation.

Children of all abilities can benefit from the study of science. Both girls and boys should be encouraged to take an active part in scientific investigations. Where necessary, we encourage children to work in single-sex groups to allow girls an equal chance to explore science. However, we are moving towards a position where girls become more confident and interested in science and single-sex groups are unnecessary. We seek to use starting points for science which appeal to girls as well as boys. In our selection of curriculum materials, we are careful to avoid stereotyping other cultures as less advanced and as the exclusive victims of natural catastrophes.

8 Resources for science.

A selection of resources is kept in each classroom. We have a right to expect these resources to be kept in good condition and to be replaced when worn. Large, occasionally used or expensive equipment is kept centrally. We are responsible for returning this equipment when we have finished using it. We all agree that the quality of our resources must be maintained so that children value the school's equipment.

9 Safety in science activities.

We accept a responsibility to plan safe activities for science. Where in doubt we refer to *Be Safe* from the ASE (see resources) or consult with the science co-ordinator. Furry mammals are not kept in the school because of the number of children with asthma.

10 Assessment policy.

Each teacher will use their assessment of the children in their class to plan appropriate scientific work. This is not left to the teachers at the end of each key stage. We try to build assessment into our teaching and do not tag it on at the end. Where short test activities are used, these should be of benefit in their own right and contribute to the children's education.

11 Record keeping.
Records of each pupil's progress are kept by their teachers and handed on at the end of the year. Class records of work done are handed on in the same way and kept in a central file.

12 The role of the head teacher.
The head teacher has a vital role in encouraging colleagues to teach effective science. He or she has responsibility for ensuring that the policy is used and for bringing the policy to the staff for periodic updating.

13 The role of the science co-ordinator.
The co-ordinator, working alongside the head, has the responsibility for progression and co-ordination of teaching the science curriculum. He or she has responsibility for the day-to-day maintenance of the science equipment and, alongside the head, for the purchase of new equipment and materials. The co-ordinator will try to support colleagues who are planning science activities. He or she will attend science courses and, as far as possible, be a resource of science knowledge for all the school.

14 Local Education Authority documentation.
The head teacher will ensure that the school's policy is congruent with national policy and that of the LEA.

Signed ....
(by all the staff responsible for drafting and agreeing this document.)

This chapter could form the basis of school INSET session in which staff review their science policy together. Each of the headings could be discussed and the staff agree to accept or modify it. The involvement of all the staff in this process means that it has the potential to become a living document of expected practice to which regular reference is made.

—20— Assessment and achievement
Shirley Clarke

This chapter outlines some of the important issues in assessment. It focuses on assessment in Britain following the introduction of the National Curriculum, but the chapter also raises issues that would be relevant in any classroom. It looks in particular at 'significant achievement' in science and how we might record this in a way which is meaningful and which will reveal what the child is learning. There are some suggested charts for use in the classroom, and the chapter could form a useful starting point for school INSET sessions.

What is assessment?

Since the introduction of the National Curriculum, assessment in Great Britain has been a subject of constant debate. Now we have the new Subject Orders within which 'levelness descriptors' replace the existing 'Statements of Attainment'. The post-Dearing climate has resulted in a situation where teachers know that endless, summative ticking of boxes, tracking every statement against every child, is not the right thing to do, but generally have no idea what this system should be replaced with.

In order to answer this question, I believe that we need to look closely at the *process* of assessment, regardless of the National Curriculum.

In the last few years, we have come to see assessment as a procedure that is done *to* the child and to believe that assessment means record-keeping. I believe that the only good reason for assessment is to enhance children's learning, to help them move forward, and to help them to become more evaluative (both in analysing things that need to be worked on and in recognising their achievements).

Assessment must be a two-way process (ultimately three-way, involving the parents) in which there is an ongoing dialogue between the teacher and the child. By this, I mean that children should know why they are doing any classroom activity and what the teacher's expectations are (e.g. 'This science investigation will help you find out more about plant growth. I also want to see how well you can design a fair test and how well you can work with Sam').

Once the purposes have been stated, the assessment agenda is set. During or after the task, in the normal course of classroom intervention, the teacher can pick up the learning intentions (e.g. 'Tell me what you think you have learnt/Do you still need more help with this?/I noticed you only worked with Sam for five minutes, then you worked separately – why was this? What do *you* think?'). This type of questioning invites the child to play an active part in her or his learning. The child can give honest opinions and can put the teacher fully in the picture (e.g. 'I understand this now/I had some help from Joe, so I think I need to do more without help/I really don't understand/Sam didn't like my idea and I didn't like his/I don't really like Sam – he's bossy').

Of course, this kind of dialogue will not take place overnight; as with problem-solving, children need to be trained and supported in the process, to learn that this kind of talk will be a constant feature and that there will be an ethos of valuing and respecting other people's points of view. As well as sharing the learning intentions with the child, it is vital that you make it clear that new and different things may be learnt which were not planned for. There is plenty of research to prove that you cannot predict what children will learn, but that you do need to have some initial focus. I particularly like Mary Jane Drummond's (1993) definition of assessment as three crucial questions that teachers must ask themselves when they set about assessing children's learning:

> What is there to see?
> How best can we understand
> what we see?
> How can we put our
> understanding to good use?

Meaningful record-keeping

If you share this view of assessment, the answers to the question 'What do I record and keep evidence of?' fit neatly into place. You focus on *significant achievement* or leaps in children's progress, which are highlighted through observation or in the assessment dialogue. Significant achievement is anything which will play a major part in the child's future achievements; not each small learning step, but breakthroughs or outstanding work. Significant achievement could consist of a leap in social skills, confidence, physical skills or understanding of concepts. Any achievement, for the first time, of any of the Statements of Attainment for Science Attainment Target 1 would indicate significant achievement, because it would affect all future science work.

Some examples of significant achievement in science are:

- setting up a fair test for the first time;
- making a prediction which, for the first time, is based on everyday experience;

- being able, for the first time, to listen and co-operate in a group investigation;
- saying, for the first time, how much the work had been enjoyed.

These are all moments which can be underplayed in a busy classroom. The child has a right, however, to have these significant moments recognised, and recorded. The child should be central in this, so that during the course of the session, when the significance occurs, the teacher in a one-to-one conversation makes much of the event (e.g. 'Well done, Sam. This is the first time you have . . . Tell me why you think that was'). The child's confidence is given a tremendous boost and has a snowball effect on future progress. A key element, at this point, is to be as clear as possible as to *why* the significance occurred, because it is this which points the teacher to future strategies for the development of that child. For example, did the leap occur because of who the child worked with, the context, the resources? The child is often the only one who knows the answer, and so must be asked.

Recording is only useful, in my opinion, if it, too, is shared with the child. The formative comments made by the teachers could be written on the child's work, in the exercise book or folder, or at least be shown to the child. Children should contribute to these comments, giving their own insights.

Significant achievement

The following list outlines the features of good formative comments which would appear on the child's actual work, or, if it is an event, on a separate piece of paper:

- date (if not already included);
- who chose this work/event as significant?;
- why is it significant?
  - physical skill (e.g. use of scissors)
  - concept (e.g. place value)
  - process skill (e.g. developing strategies to solve problems)
  - social skill (e.g. turn-taking)
  - attitude (e.g. increased confidence).

The latter point should be *positive* significance, and, when appropriate, should refer to what was happening before (e.g. previously only watched others at play, now can join in). The significance should be explicit (i.e. the first time . . . and mention should be made of:

- context and level of support (if it affected the significance);
- the child's opinions and comments;
- next steps for the child;
- references to the National Curriculum, if appropriate.

For example:

> Date: 17/3/94
> This is the first time that Amy chose to be the scribe to write out a science investigation carried out by her and Claire. She said, 'Claire was a bit shy and so I said I would do it'. She also said she thought her work was good. Amy needs to work with other similar children.

A key point for the teacher is the reason for this significant leap. In the example above, the teacher had previously grouped quieter children with more confident children. The pairing of two less confident girls enabled Amy to express herself. Clearly she would benefit from more pairing with similar children.

Significant progress links directly to the evidence of the accompanying work. The work can be photocopied or cut out of exercise books and kept in a Record of Achievement, which again should belong to the child. Then, when the next leap is evident, the previous work can be taken out and progress determined (e.g. 'Let's look at the investigation you did last month and compare it to this one . . .'). This is a highly motivating process in which children see directly the progress they have made. At this point, further aims or targets are discussed with the child (e.g. 'You need to work at . . .'). In this way, the folder of evidence is used purposefully.

Recording progress

Up to now, many schools have kept huge evidence folders of work, which are not shown to the child and which cause many teachers to ask, 'Why am I doing this? What use is it, especially as it becomes redundant very soon after it is collected? Where will I store it, and who will look at it?'. Significant evidence can also be transferred to the next teacher, after sifting it down to just a few pieces of work. For this purpose, it may be useful to add more detail to the annotations, providing information about the child which the next teacher needs.

Assessing children's science effectively depends on organising the children to work so that the teacher can be another resource; having ongoing dialogues with children about their work, either when they finish an activity or when they are in the middle of it. If the children are aware of the learning intentions of the activity, the dialogue can focus on those intended outcomes and on other, unexpected outcomes which may have arisen from the processes the children were using. Two ways in which I have seen significant progress being recorded are as follows:

1 Each child has an ongoing task sheet with columns headed (Figure 20.1). The teacher writes those tasks which produce rich information (e.g. investigative work) on the sheet, either in advance or as the child

Figure 20.1   Record sheet 1.

Child's Name:

Activity	Learning Intentions	Date (started/repeated/ended, if appropriate)	Comment
	↑ SOAs and other intentions		↑ by teacher 'on the hoof' sharing comments with child and inviting child contribution
↑ Science investigations for two weeks or more (usually those with most potential for significant progress)			

Figure 20.2   Record sheet 2.

ACTIVITY ↑ science investigations for a half term	Learning Intentions	Children's Names
	a) b) c) d) ↑ could be SOAs, or other, such as cooperative skills	Note where significant attainment occurs – a tracking system, which leads to the evidence

starts, and writes three or four learning intentions in the second column. The children keep these sheets beside them as they work, making them accessible to the teacher as she or he is engaging in dialogue with them. With pen in hand, it is then relatively painless to jot down comments relating to the learning intentions and including points made by the child. Notes are only made when there is significant achievement; otherwise the system could get in the way of teaching and learning. The sheets can be used for transferring onto summative sheets, and for passing on to the next teacher. The drawback, however, could be that some children will fall through the net.

2 The teacher has a large matrix (to last for a week, a fortnight or more) with all the children's names along the top, and space for appropriate 'rich' activities and learning intentions down the side (Figure 20.2). Only some of the children would be involved in each activity, so the matrix does not link all the children to each activity. The process of recording is then the same as in the first system. The actual assessment takes place during the teacher's ongoing dialogue with the children; but significant progress is marked onto the matrix, with the formative comments written on the child's actual work or on the page in the child's book where the task has been written. This system allows for all the children's progress to be viewed at once, and thus children who have not apparently made significant progress can be focused on specifically.

Conclusions

It is a common misconception that a good assessment system will ensure good learning. All that a system can do is structure and organise the assessment process and recording of assessment outcomes; it is of no use to anyone if the assessment system is perfect but the children are underachieving and the information is not acted on. Assessment is concerned with understanding children's understanding. This involves knowing how best to question children, when to intervene, when to let children use invented methods, when to teach them something new, how to make the most of science starting points and how to help children become successful, confident scientists. After three or four years of emphasis on assessment, isn't it about time we looked at how children best learn science and how to be a good science teacher? Without the wealth of science advisors we used to have, there are fewer chances these days for teachers to receive good-quality INSET.

My advice to any school embarking on a review of their current science assessment systems would be to look first at how children learn, how you can best facilitate their learning, what you want children to learn and how you can provide for this while ensuring differentiation – and then think about how to build in assessment systems.

—21— Getting better at investigations
Robin Smith

This chapter pulls together some of the themes in Section C about children investigating

> THEME: helping pupils develop their skills and investigations in classrooms

Introduction

In primary schools, there is a long tradition of children exploring their own ideas and finding answers to their own questions. Recently this has been developed into a more specific focus on scientific investigations. In England and Wales, the introduction of Attainment Target 1 (Science 1) in the National Curriculum accelerated this, but also in some ways complicated it because of the jargon that was introduced and the accompanying pressures of assessment. Although the changes now imminent in Science 1 will alter details, the main thrust remains the same; in any case, the task is one that concerns teachers whether or not they have a national curriculum: how can we help children become better investigators through their primary years?

What does getting better mean? It means they become more able to plan and systematically carry out their own, whole scientific investigations. As teachers, we need to get better at supporting this aspect of their science learning. To do that, we first need to feel at home with what scientific investigation involves and how to make it happen in busy classrooms.

What do we mean by 'investigation'?

There are two ways of looking at pupils' scientific investigations, it seems to me. Both are necessary. The first is to consider them as the extension of what children do naturally. In this view, infants explore the world and experiment to find out how things behave and what will happen if . . .

Parents and teachers of young children are probably only too familiar with this urge to examine and act on the world. Jos Eltgeest (1985) wrote vividly about 'Tiny Niels' investigating the effect of waves on the beach.

> He could not talk yet, not a word was exchanged, no question formulated, but the boy himself was the questions, a living query: 'What is this? What does it do? How does it feel?'.

Closer to home, the National Curriculum requirement that pupils 'should experience the natural force of gravity pulling things down' would seem to be actively explored at an earlier stage by most infants when they drop things out of their pram for willing adults to pick up – again and again.

Running through this view is the theme of children, prompted by inner questions, actively learning through their search for reasons or results. Sadly, some children sooner or later lose much of the curiosity or confidence to risk such exploration of their ideas. Is it the media, maturation or the pressure of home and schooling which restricts them? Whatever the cause, our task is often to sustain and refine such curiosity so it can be used for personal development as well as to serve the narrower needs of teaching science.

A second view is from the perspective of scientific activity. What is it that scientists do when they investigate and how far is that appropriate in schools? This model seems logical but risks oversimplifying, and idealising, the untidy and varied activities of research scientists. However, it has provided a more detailed picture of what we can look for and work toward with our pupils. We should not let the jargon and structure of the versions laid down for us in curriculum proposals get in the way of our appreciation of the power and composition of this picture. The details include features with which primary teachers have long been concerned, such as careful observation, and others with which they are becoming familiar, like the identification and control of variables to make for 'fairer tests'. Further elements which may take longer for us to build into our daily practice include the interpretation of results, the evaluation and refinement of experiments.

The three strands currently distinguished in 'AT1' or 'Science 1' are represented by those examples above: asking questions in increasingly informed ways; observing and managing variables; interpreting their findings. The ideal held up by the curriculum is for pupils to combine these in whole investigations which they have designed to answer their own questions – science with reason.

If the goal is for our pupils to be able to do whole investigations and become more autonomous, there are lots of small steps on the way. Let's look at some examples at different stages of that journey.

Getting started

How should we begin when pupils, or their teacher, are not very accustomed to an investigative approach?

Limit the variables

The variables are the things that can be changed and controlled in any investigation. For instance, if your pupils are curious about the factors influencing germination of seeds, they will need help to investigate whether light really is crucial, and you may need to ensure other variables, such as temperature or water, are controlled and, if appropriate, examined separately.

Use familiar materials

Standard experiments in books sometimes use contexts and equipment that derive from adult or secondary school experience rather than the world of young children. An investigation of materials to see which is best for insulating against heat loss is more likely to be meaningful if, instead of beakers or even yoghurt pots, the warm water is in 'clothed people' modelled around washing-up liquid bottles, or in hot-water bottles; in summer, there might be even more reason for children to learn fast if they are trying to slow down the melting of their ice-lollies!

Focus on one aspect or skill at a time

This can be done by raising questions or checking and repeating observations, but not both. Thus if your priority is to get children to ask relevant questions about the toy vehicles they have been rolling in play, you will encourage them to offer as many questions as they can and to discuss how they might be tested. However, if your focus is on improving their observations, you might ensure they roll cars in such a way that there is a need to look closely and repeatedly at where they run.

Structure and support

Provide structure and support for other aspects of the investigation so that the demands of the task do not get in the way of its main purpose. Problems in working with others and in recording results are often the preoccupation of pupils who really should be concentrating upon improving their test, for instance of how well different paper towels soak up water. Careful grouping and provision of a simple recording sheet may free them to focus on the main task. This is a case where more structure can, in fact, provide more freedom to learn.

Pre-teach skills

When appropriate, teach skills the children will use, such as correct use of a microscope or measuring vessels, but make sure they see the need to learn this and have a reason for applying it. Once they know how to use apparatus, even the youngest children are able to investigate more independently, provided the resources are available when needed. Many infant teachers are especially skilled at building the teaching of new skills into their integrated topics.

Whole-class activities

Some of this work is best done with small groups but there are many class activities which can be used to foster more independent investigation. Class discussions can be particularly valuable at the start to provoke predictions and hypotheses, for instance about how to get shadows of a required size in a puppet theatre. It is just as important to find time after children have explored their ideas for them to share their findings, as this establishes the right climate and gets the process of interpreting and evaluating started.

Getting better

What should we do to improve the way investigations are done by our pupils?

Build on the skills and insights they have already

So if they have learnt how to organise themselves, can safely get equipment out and return it, and know how to weigh, then they could be encouraged to plan and carry out an investigation to find how people check the loads in the barges they saw on their visit to the canal.

Identify, teach and assess any new skills they will need

These could include using a spring balance, or making tables of results and working out averages. Don't be afraid of leading and teaching, but be clear about the purpose of practical work: don't confuse investigation with practicals which can legitimately be done for a different purpose, for instance to demonstrate a technique or to illustrate a phenomenon.

Help them evaluate their own plans and improve their investigations

Aids such as an experiment-planning sheet can be very useful at the right stage (Barnett and Green, 1993). Introducing pupils to different ways of recording results as they go through the school can lead to them eventually being able to choose the best way to communicate a particular set of results.

Get a bit more demanding

Give more open challenges and increasing responsibility as pupils gain in skill and confidence. Seize opportunities and encourage adventurous ideas. However, be ready to provide a safety net to avoid discouragement – be prepared with some ideas you have tried before so you know they will work. Take small steps and succeed – children and their teachers deserve success.

The chapters in Section D have described approaches to general issues that arise in teaching science. They are issues which cannot be ignored and appear in much of the earlier work. We hope these chapters have provided starting points for individual teachers, co-ordinators or schools to address them.

22 — Conclusions
Sue Atkinson

This book has attempted to bring out some of the features of 'good practice' in teaching science. We have tried to link it to current theories about how children learn and have tried also to show how we think that children might learn in similar ways in different areas of the curriculum.

The viewpoint we have taken in this book could be called 'constructivist'. In pulling these threads together, it might be helpful to show some elements of what we might mean by 'constructivism'.

What does it mean in my classroom to have a constructivist approach?

For the learner (and this includes the teacher, as he or she is always a learner)

- Learning is based on purposeful activity.
- This active process is the means by which learners construct their own meaning for themselves, often with someone else.
- Learners are seen to be responsible for their own learning.
- Learners bring what they know to the learning situation, and this is valued and respected.
- This active learning and construction of meaning involves many processes, inevitably talking, listening and questioning, usually some kind of group negotiation, and often some kind of problem solving.
- Knowledge for the learner is something personal. The learner will test what they learn and observe to see if it fits into what they already know.
- The learner is given time and space to reflect on their experience.

For the teacher

- Learners are viewed as active, not passive.
- The views that learners hold at the outset of a piece of work must first be explored in order to build on what they know.
- Teaching will be seen less as the transference of knowledge and more about the organisation of situations in the classroom which enable learning to take place.

- Teachers also come to the learning situation with their own knowledge, understandings and experience, and this influences the ways in which they respond to the learners, the tasks they will expect them to do, their roles as the tasks are carried out, the types of outcomes they expect and aim for, and their own professional development.

- The curriculum is seen as something much more than a list of content to be covered. It is seen as an active set of tasks and equipment that the teacher organises in the learning environment in order to facilitate the activity of the learners, enabling them to construct their own meaning for themselves.

- The teacher's view of knowledge may be that all knowledge is problematic in nature and socially constructed, but this need not be so. It is possible to believe that taking a constructivist standpoint in the classroom aids learning, even though the teacher may believe that there are 'facts' independent of our knowing them.

- We might see ourselves as helping the learner appreciate 'standard scientific' viewpoints.

- We reflect on what we learn in our teaching and try to reflect on this.

Many teachers say that they feel much more confident with teaching language than science, and we have tried to show that, although science teaching has its own special difficulties (I think I will spend the rest of my life finding energy and forces difficult to understand!), we can use our strengths and skills from other areas and apply them in science teaching. Developing our ability to ask open-ended questions, for example, or getting our children to say exactly what they mean are things that are crucial in any curricular area.

There are many books available now to help us with the problem of understanding the content of science. I now use my own children's secondary-school books. In the resources section we have listed some recent publications designed to help teachers with this problem.

The stories in this book have shown the crucial role of talking, listening, questioning and reasoning in teaching science. This talking, along with children's own recordings, can often be the key to effective science teaching in that we can start to see what it is that the children are thinking and what they know. If we make our starting point finding out what they know and ask them what they want to find out next, we can work alongside the children, supporting them as they grope their way towards the more 'standard scientific' viewpoints. What makes science difficult is that it is less like the language you already know. For example, scientists' use of the word 'energy' is not the same as the everyday one. Science has its own distinctive ways of talking, but that does not make them exceptionally difficult – just unfamiliar.

Teaching science is not an easy process, but we hope that you have felt some of the excitement that the teachers demonstrate in their stories and that you will want to have a go and challenge yourself to go on exploring science with reason.

References and Resources

References

ATHEY, C. (1990) *Extending Thought in Young Children: a parent–teacher partnership.* London: Paul Chapman Publishing Ltd.

ATKINSON, S. (1992) *Mathematics with Reason.* London: Hodder and Stoughton.

ASUBEL, D. (1968) *Educational Psychology.* New York: Holt, Rinehart and Winston.

BARNETT, R. and GREE, D. (eds) (1993) *Key Stage 2 Atlas Resources Sc. 1 Scientific Investigations.* Sheffield Hallam University, Centre for Science Education.

Bayer Periodic Table of the Elements. BAYER AUSTRALIA LTD., Botany, NSW. Available from the South Australian Science Teachers' Association, 163A Greenhill Road, Parkside, S.A. 5063 Australia.

BEARLIN, M. (1990) 'Towards a gender-sensitive model of science teacher education for women primary and early childhood teachers.' *Research in Science Education* Vol 20 pp. 21–30.

HARDY, T., BEARLIN, M. and KIRKWOOD, V. (1990) 'Outcomes of the primary and early childhood science and technology education project at the university of Canberra.' *Research in Science Education* Vol 20 pp. 142–151.

BIDDULPH, F. and OSBORNE, R.M. (eds) (1984) *Making Sense of Our World: An Interactive Teaching Approach.* Hamilton, NZ: Science Education Research Unit, University of Waikato, Hamilton, New Zealand.

BISSEX, G. (1980) *Gyns at Work: a child learns to read and write.* Cambridge Mass: Harvard University Press.

BLISS, J., OGBORN, J. *et. al.* (1989) 'Secondary pupil's common-sense theories of motion.' *International Journal of Science Education* 11 (3): 261–272.

BRUNER, J. (1987) The transactional self. In J. Bruner and H. Haste (eds) *Making Sense:* The child's construction of the world. New York: Methuen.

BRYSON, J., MATSON, M. and VEZZUTO, L. (1986) *The Anatomy Apron.* No. 2534M, Dominguez Hills, CA, Educational Insights.

BURNINGHAM, J. (1989) *Oi! Get Off Our Train.* London: Walker Books.

CAREY, S. (1985) *Conceptual Change in Childhood.* Cambridge, Mass: MIT Press.

CARRUTHERS, E. (work in progress) 'Looking at aspects of developmental literacy and exploring this concept in mathematics.' Unpublished M.Ed. thesis, Plymouth University.

DAVID, T. (1990) *Under Five – Under-educated?* Milton Keynes: Open University Press.

DES/WO *(1991) Science in the national curriculum.* London: HMSO.

DEWEY, J. *(1916) Democracy and Education.* London: Macmillan.

DI SESSA, A.A. (1988) 'Knowledge in Pieces'. *Constructivism in the Computer Age.* Hillsdale, New Jersey: Lawrence Erlbaum Associates.

DONALDSON, M. (1978) *Children's Minds.* London: Fontana.

DRIVER, R. (1983) *The Pupil as Scientist?* Milton Keynes: Open University Press.

DRIVER, R., GUESNE, E. and TIBERGHEIN, A. (1985) *Children's Ideas in Science.* Milton Keynes: Open University Press.

DRUMMOND, M.J. (1993) *Assessing Children's Learning.* London: David Fulton.

EARLY YEARS CURRICULUM GROUP (1989) *Early Childhood Education:* the early years curriculum and the National Curriculum. Stoke-on-Trent: Trentham Books.

EDELAM, G. (1992) *Bright Air, Brilliant Fire.* London: Penguin.

ELTGEEST, J. (1985) in Harlen W. (ed) Primary science: taking the plunge. London: Heinemann.

FARRAR, R. and RICHMOND, J. (eds) (1979) *National Oracy Project how talking is thinking.* London: ILEA learning materials service.

FLEER, M. (1991) 'Towards a theory of scaffolded early childhood science education.' *Australian Journal of Early Childhood* 16(3), 14–22.

FLEER, M. and LESLIE, C. (1992), *The Light Side of Darkness.* AECA resource booklet.

GOODMAN, K. (1986) *What's whole in whole language?* Ontario: Scholastic Educational.

JAMES, S. (1991) *Dear Greenpeace.* London: Walker books.

KELLY, A. (ed) (1987) *Science for Girls?* Milton Keynes: Open University Press.

LALLJEE, B., STICKLAND, R. and HOLDERNESS, J. (in press) *Frameworks for Talk: speaking and listening in the primary school.* London: Simon and Schuster.

LIPMAN, M. (1980) *Philosophy in the Classroom.* Temple University Press.

LIPMAN, M. (1988) *Philosophy Goes to School.* Temple University Press.

LIPMAN, M. (1991) *Thinking in Education.* Cambridge UK: Cambridge University Press.

LIPMAN, M. (first published in 1974, but now abridged and adapted for use in Britain by Sutcliffe, R. 1993) *Harry Stottlemeier's Discovery.* (Published privately, see Resources section.)

MARTIN, J. (1990) 'Literacy in science: Learning to handle text as technology', in Christie, F. (ed) *Literacy for a Changing World.* Australian Council for Educational Research: Australia.

MURRIS, K. (1992) *Teaching Philosophy with Picture Books.* London: Infonet Publications.

OSBORNE, R.J. and FREYBERG, P. (1985) (eds) Learning in Science: The Implications of Children's Science. Auckland: Heinemann.

Primary and early childhood science and technology course, information available from Dr. Tim Hardy, Faculty of Education, University of Canberra, PO Box 1, Belconnen, ACT 2616, Australia.

RAYNER, C. (1978) *The Body Book.* London: Pan Books Ltd.

ROGERS, C. (1983) *Freedom to Learn.* Toronto: Macmillan.

ROYSTON, A. (1992) *What's Inside? My Body.* London: Dorling Kindersley; Australia: RD Press.

SAPERE The society for the advancement of philosophical enquiry and reflection in education (see Resources).

STICKLAND, R. (in press) 'Talk across the curriculum' in Lalljee, B. *et. al. Frameworks for talk: speaking and listening in the primary school.* London: Simon and Schuster.

SPACE Science Process and Concept Exploration Project: This was a joint research project between the Centre for Educational Studies, King's College and the Centre for Research in Primary Science and Technology, Liverpool University from 1988–1992. So far eight research reports – Light, Electricity, Processes of Life, the Earth in Space, Growth, Evaporation and Condensation, Sound, Materials and Rocks, and Weather – have been produced. Copies of the reports are available from Liverpool University Press, PO Box 147, Liverpool L69 3BX.

Time Life Periodic Table of Elements. Available from Charles Fogliani, Applied Science, Charles Sturt University, Bathurs NSW 2795, Australia.

TIZARD, B. and HUGHES, M. (1984) *Young Children Learning: Talking and Thinking at Home and at School.* London: Fontana.

VYGOTSKY, L. (1978) *Mind in Society.* Cambridge, Mass: Harvard University Press.

WERTCH, J.V. (1985) *Vygotsky and the Social Formation of Mind.* Cambridge, Mass: Harvard University Press.

WOOD, D. (1986) 'Aspects of teaching and learning', in Richards, M. and Light, P. (eds) *Children of Social Worlds.* Cambridge: Polity Press.

WOOD, D. (1988) *How Children Think and Learn.* Oxford: Basil Blackwell.

Resources

Basic information

Sherringham, R. (ed.) (1993) *ASE Primary Science Teachers' Handbook*. London: Simon and Schuster.

Starting from thinking and talking

If you are interested in exploring science from talking and reasoning, then Roger Sutcliffe's book is a good place to start. (Sutcliffe, 1993) *Harry Stottlemeier's Discovery*. Book available from Roger Sutcliffe SAPERE. This book is one of a number of specially written stories which include practical support for teachers. Details of this book and others from R. Sutcliffe, Stammerham North, Christ's Hospital, Horsham, RH13 7NF.

Fisher, R. (1990) *Teaching Children to Think*. Oxford: Basil Blackwell.

Science content

If you want help with the content of science these might help.

The *'Understanding science concepts teacher education materials for primary school science'* series. There are six titles.
Understanding Forces; Understanding Energy; Understanding Living Things; Understanding Materials and why they Change; Understanding the Earth's Place in the Universe; Understanding Floating and Sinking.

All of these are available from ASE: Association for Science Education, College Lane, Hatfield, Herts. AL10 9AA. Tel. 01707 267411.

Nuffield Primary Science, based on SPACE project   If you have liked the approach to science that we have taken in this book, but want more support in your classroom, you might like the Nuffield Primary Science materials (Collins 1993).

They have been produced as a result of the SPACE project. These resources are divided into key stages 1 and 2 and are a series of attractive pupil booklets (e.g. 'Sound and music', 'Rocks, soils and weather' and 'Where things live') full of ideas, starting points and information. Each pupil booklet is accompanied by a well-planned teacher's book.

These Teacher's Guides have references to children's ideas and examples of assessment and notes on the knowledge that the teacher needs. There is a very good Teachers' Handbook that explains some of the approaches to science learning advocated by the SPACE project. This is easy to read, informative, and has relevant information such as assessment ideas and other essential reading if you want to use the resources in the best possible way.

SATIS 8–14   Good resources for children over eight. These materials are photocopiable resources rather than a course and as such are flexible and teacher and child friendly. They address a wide range of real and relevant problems in society (waste disposal, building new roads, wildlife, health issues) as well as really good starting points for design and technology in the classroom. The breadth of the materials is enormous and there is plenty of information for the non-specialist.

Collins Primary Science   This is divided into key stages 1 and 2 and is a series of well produced, attractive and colourful books for children packed with ideas for investigations, (e.g. 'Clothes', 'Fruit and vegetables', 'My school'). These books could be used by children as starting points. and they make good books for the reading corner. There are also assessment books that are easy to use and realistic about teacher time, and some books for children to work in called 'my record book for science' in which the child, parent and teacher work together to support the investigations.

Science and Technology in Society, 1992   Produced as a result of the Science and Technology in society project. Published by ASE (address above).

Ginn Science Reception Games   These games provide preparatory work for children of four and five years of age before they start out on science in year 1.

Peacock, G. and Smith, R. (1992) *Teaching and Understanding Science*. London: Hodder and Stoughton.

Science and gender

There is a number of books in the area of gender issues in science, maths and technology.

Browne, N. (ed.) (1991) *Science and Technology in the Early Years*. Milton Keynes: Open University Press.

Fensham, P. (ed.) (1988) *Development and Dilemmas in Science Education*. London: The Falmer Press.

Harding, J. (ed.) (1986) *Perspectives on Gender and Science*. London: The Falmer Press.

Kelly, A. (1981) *The Missing Half: Girls and Science Education.* Manchester: Manchester University Press.

Titchell, E. (ed.) *Dolls and Dungarees.* Milton Keynes: Open University Press.

Keeping a multicultural perspective

Materials from Global Education Projects, Global Futures Centre, Maenllwyd, Llangynog, Carmarthen, Dyfed, SA33 55A. Materials include, Rainforest Child, Tomorrow's Woods, An Arctic Child, etc.

Fisher, S. and Hicks, D. (1985) *World Studies 8–13: A Teacher's Handbook.* London: Oliver and Boyd.

Assessment

Esme Glavert (in press) *Tracking Significant Achievement in Primary Science.* London: Hodder and Stoughton.

Other suggestions

The Early Start books published by Simon and Schuster, for example Richards, R. (1989) *An Early Start to Nature.*

Safety note

Many science investigations have hidden dangers.

- Consult any guidelines produced by your local education authority or area science advisory team.
- Be aware of all the details of safety needs by reading *Be safe! some aspects of safety in school science and technology for key stages 1 and 2* (2nd edition, 1990) from ASE address above.

Index

assessment 160–5
Asubel, D. 135

Biddulph and Osborne 7, 118
Bliss, J. 23
body 19–20, 48–57
bones 115–24
bridges 75–81
building site 90–9

chemistry (see elements)
classroom organisation 42–6, 148–54
community of enquiry 32–41
concept map 40, 48–57, 58–66, 68, 83, 102, 131, 137, 149
constructivism ix, 170–1

discovery approach 4
Driver, R. viii, 39

early years 9, 13, 32–89, 148–54
electricity 34, 82–9
elements 125–34
energy 107–14
environmental issues 58–66, 107–14, 153

fair test 26–31, 75–81, 90–9, 135–46
flight 26–31
forces 26–31, 75–81

geology 135–46
Goodman, K. 14

interactive approach 7, 48–57, 58–66, 67–74, 82–9, 100–6, 107–14, 115–24, 125–34
intuitive viii–x, 15–24
investigations, throughout, especially 26–31, 70, 86, 90–9, 115–24, 130, 148–54, 166–71

juniors 15–24, 90–154

language (see oracy and talking)
learning process vii, 2–8, 15–24
light 17–19
Lipman, M. 33–8

mind map (see concept map)
movement 67–74

national curriculum throughout especially 160–5
new knowledge 39

open/closed questions 43
oracy 32, 42–6
outer space 100–6

paper planes 26–31
periodic table 129
philosophy 32–41
Piaget, J. 19
plan, do, review 26–31, 90–9
planet 107–14
planning 27
plants 58–66, 107–14, 148–54
policy, school 155–9
pre-school 48–74
process approach 6, 90–9
physics 15–24, 26–31, 67–106

questioning 32–46, 67–74, 84–5, 91, 93, 97, 103, 120, 136, 140, 148

recording 29, 131, 148–54, 160–5
resources 176
rocks 135–46

SAPERE 32, 176
school policy 155–9
science at home 9–14
science notice board 26–31, 90–9, 148–54
significant achievement 162
space (see outer space)
SPACE project viii, 15–24
structures 75–81
symbols 66, 125–34

talking 21–3, 32–41, 42–6, 97, 106, 171
teacher interventions 32–41, 75–81, 90–9, 115, 154
teaching strategies 26–46
telephones 82–9
transmission approach 5

universe 125–34

Vygotsky, L. 32, 77